HOW
YOUR BODY
WORKS

SUSAN L. ENGEL-ARIELI, M.D.

Ziff-Davis Press
Emeryville, California

Development Editors	Mary Johnson and Cheryl Holzaepfel
Copy Editor	Mary Johnson
Technical Reviewer	Bryce Palchick, M.D.
Proofreader	Carol Burbo
Cover Design	Regan Honda
Book Design	Carrie English
Digital Illustration	Dave Feasey and Sarah Ishida
Word Processing	Howard Blechman
Page Layout	Bruce Lundquist
Indexer	Valerie Robbins

Ziff-Davis Press books are produced on a Macintosh computer system with the following applications: FrameMaker®, Microsoft® Word, QuarkXPress®, Adobe Illustrator®, Adobe Photoshop®, Adobe Streamline™, MacLink®*Plus*, Aldus® FreeHand™, Collage Plus™.

If you have comments or questions or would like to receive a free catalog, call or write:
Ziff-Davis Press
5903 Christie Avenue
Emeryville, CA 94608
1-800-688-0448

ISBN 1-56276-231-1

Manufactured in the United States of America
✪ This book is printed on paper that contains 50% total recycled fiber of which 20% is de-inked postconsumer fiber.
10 9 8 7 6 5 4 3 2 1

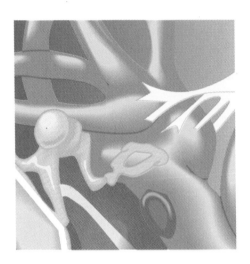

This book is dedicated to my mother,
Marion Carpenter Kasper, for her
herculean efforts in helping me attain
many of my goals in medicine, and for
her unconditional love and support.

Introduction ....**xi**

Chapter 1
The Brain and
Nervous System........................**1**

Chapter 2
The Eye...................................**14**

Chapter 3
The Ear **24**

Chapter 4
Your Nose and Senses
of Smell and Taste **32**

Chapter 5
The Skin, Hair, and Nails **40**

Chapter 6
The Bones.............................**48**

Chapter 7
The Muscles **58**

Chapter 8
The Arms, Forearms, and
Hands **66**

Chapter 9
The Legs and Feet **76**

Chapter 10
The Heart and Circulation....**86**

Chapter 11
The Lungs...............................**98**

Chapter 12
The Stomach, Intestines,
Gallbladder, and Liver **108**

Chapter 13
The Kidneys and the
Urinary Tract System **120**

Chapter 14
The Thyroid, Pancreas, and
Other Glands **126**

Chapter 15
The Male and Female
Genitalia **136**

Chapter 16
Cells, Pregnancy, and the
Formation of a New Life **148**

Index....................................**159**

No woman or man stands alone in the production of an undertaking of this magnitude. This book was molded by many hands and minds. Therefore, I would like to express my deepest gratitude to the many people who have journeyed down the path with me, and helped in the production of this book.

First, I would like to thank Cindy Hudson, President of Ziff-Davis Press, for her encouragement and motivation in this book. Eric Stone, Acquisitions Editor, was also extremely helpful, understanding, and was always available to answer questions and concerns.

My heartfelt thanks go out to my editors (in the order of coming onto the project): Melinda Levine for her exacting attention to details and asking questions, and Sharon Maddox for her editorial style and flair. Cheryl Holzaepfel, Editor-in-Chief, was involved every step of the way, and I am grateful for, and thank her for her help with the illustration concepts and more. Mary Johnson, who participated in the majority of the editorial assistance for the book, is also thanked for providing editorial guidance and questions and ideas regarding materials to include in the book. Last, but not least, my gratitude goes to Ami Knox, Project Coordinator, for her perseverance, skills, and patience in the revision process.

Dr. Bryce Palchick, of Pittsburgh, Pennsylvania, a reviewer for this book, was an invaluable source of ideas, information, and corrections, particularly since my deadlines entailed that the writing had to be done at race car speed. This book was enriched by his input, and he is appreciated and greatly thanked.

Immense thanks also go to the computer art illustrators at Ziff-Davis Press, Dave Feasey and Sarah Ishida, for their computer graphic masterpieces. This was no small feat, and I congratulate and thank all the illustrators very much.

I am also very grateful to many other people at Ziff-Davis Press for their help behind the scenes. They include, in alphabetical order: Howard Blechman (for word processing), Carol Burbo (for proofreading), Bruce Lundquist (for page layout), and Cori Pansarasa (for proofreading and editorial assistance).

Much appreciation goes to Lutheran General Hospital and Dr. John P. Anastos, D.O., radiologist at Lutheran General Hospital for his radiologic interpretations and radiologic films for this book. Additional thanks go to GE Medical Systems and Peg Ireland and Brian Johnson of GE Medical Systems, Milwaukee, Wisconsin for providing additional radiologic films; and to Jim Gardner, VP Marketing of Neuro-Com International of Clackamas, Oregon for his assistance and photographs.

Gratitude also goes to Bill Kasper for his patience and encouragement, and for helping me with related tasks in researching and preparing this book.

Lastly, I would like to thank my husband, Udi Arieli, for his great computer assistance and critiques for this book; for his patience during this undertaking; and for his loving support during the writing of this book.

The human body was created for the sake of the soul and stands erect among other animals....
Man alone has the power of thought; likewise he alone remembers.

—Alessandro Benedetti (1450–1512),
The History of the Human Body

This book will take you on a wonderful journey through your body. It was written for anyone with a desire or need to know more about the body and how it works. If you struggled through high school biology, or if you find the body puzzling and complicated, this book is for you. If you want to learn more about this wondrous machine and how to keep it healthy, this book is for you, too. If you (or someone you know) have a health problem and want to learn more about it, this book can also help you.

People naturally crave knowledge about the body. A look at the mechanics of the human body provides a complex and beautiful panorama—but we are still a long way from completely understanding what we see. Each of us contains a magical combination of cells that forms tissues and organs and that allows us to function in the world around us. Imagine for a moment all that is going on in your body right now! As your eyes take in the images on this page, electrical impulses are being sent to your brain to interpret the information; your hand muscles are preparing to turn the page; your heart is beating; your lungs are expanding; new blood cells are being produced; billions of other cells are being created, living, or being made extinct.

This book contains 16 chapters about the brain, eyes, ears, nose, tongue, heart, lungs, bones, muscles, arms, legs, abdomen, glands, and sex organs and about pregnancy and other topics. The material is presented in simple language and with an abundance of illustrations that clearly show you how your body works. Although far from comprehensive, this book will provide you what you need for a fundamental—and practical—understanding.

In addition to showing the workings of your body, each chapter also discusses some common maladies, with accompanying illustrations. A sampling of the topics includes tension and migraine headaches, nearsightedness, skin cancers, osteoporosis, ankle sprains, stomach ulcers, the effects of smoking and excessive alcohol consumption, diabetes, breast cancer, prostate cancer, and other common problems. Again, the material is not all-encompassing—it's not meant to enable you to diagnose an illness (that should be left up to your own healthcare professional, of course). What it will provide is a sense of the magnitude, symptoms, and workings of a disease or condition. Modes of treatment have not been included as they are beyond the scope of this book; new discoveries are constantly being made, and no two illnesses are exactly the same or use exactly the same treatments.

It is the author's hope that you will find this book useful and interesting, for that is what we have all strived to obtain. We hope to have answered many of your questions—and to have brought forth some new ones, as well.

CHAPTER 1

The Brain and Nervous System

THE NERVOUS SYSTEM consists of the brain, spinal cord, peripheral nerves emanating from the spinal cord, and the autonomic nervous system, which controls involuntary functions in the body. The brain and nervous system together provide most of the master controls of the body and regulate thinking, sleeping, perception of pain, breathing, muscle contractions, secretion of some glands, temperature, the arousals and sensations of sex, and more. Furthermore, they process and store information.

One of the major functions of the brain and nervous system is to process incoming information in such a way that appropriate responses occur. Most of the information received by the brain is discarded as insignificant. If it did not do this, you would be bothered by the pressure of clothes on your body and disturbed by all the objects in your field of vision and all the noises around you. Also, only a small amount of important incoming information produces an immediate response. Most information is stored in the cerebral cortex of the brain as memory. Once memories are stored, they become part of a processing chain that channels and integrates them into new thinking information.

The brain has three major components: the cerebrum, cerebellum, and brain stem. The cerebrum receives information, thinks, processes, and sends out information. The cerebellum is essential for good balance and coordination. The brain stem connects the cerebrum to the spinal cord. The brain stem is very complex anatomically, but it primarily consists of the medulla oblongata, which regulates functions such as respiration, heart rate, and blood flow; the pons, which controls eye movements and pupil size; and the midbrain, which controls eye, neck, and head movements.

Sleep is indispensable to our existence and is controlled within the medulla, pons, and other areas of the brain such as the hypothalamus. Our sleep-wake cycle appears to be age related. Each day, a newborn baby sleeps between 16 and 20 hours; a teenager sleeps about 7.5 hours, and an older adult, about 6.5 hours. We know that there are five stages of sleep, based on studies of brain waves using a test called an EEG (electroencephalogram) and on clinical studies. When you first begin to sleep, your muscles relax and eye movements become slow and rolling in nature; this is known as stage 1 sleep. Stages 2, 3, and 4 continue to produce deeper levels of sleep; stage 4 is the deepest. The fifth stage of sleep, called REM (rapid eye movement), is where complex visual dreaming occurs.

Pain is also regulated by the brain. Different tissues in your body have different responses to injury. Pain receptors on the skin respond to most types of skin injury, such as pricking, cutting, and crushing. However, these same stimuli produce no pain in other tissues such as the stomach, intestines, joints, or brain. Instead, pain in the stomach, intestines, or joints is induced by distention, spasm, or inflammation. Pain in the head is produced from distention or inflammation in the accompanying blood vessels or nerves.

The brain controls, and is essential for, the regulation of breathing mechanisms. The respiratory center is located in the medulla oblongata and pons within the brain stem. Inspiratory and expiratory nerves, which control inhalation and exhalation, intermingle in this area. High concentrations of carbon dioxide, and, to a lesser extent, high acid content in the blood stream stimulate your taking a breath to draw more oxygen into the body. Increased levels of carbon dioxide in the blood, detected by the respiratory center of the brain, stimulates the inspiratory nerves and inhibits the expiratory nerves, which sends signals to the chest to increase inspiration and breathe. Additional receptors in the carotid arteries in the neck and in the aorta in the chest also help in regulating breathing. When the carbon dioxide content is sufficiently decreased, the inspiratory nerves are inhibited, then the expiratory nerves are freed from inhibitory signals, which creates expiration, and there is a release of carbon dioxide from the body.

Diseases and malfunctions of the brain have the potential to cause more damage and suffering than those that occur in many other parts of the body. Of all the painful states that afflict us, headaches are the most frequent. The brain tissue itself is almost insensitive to pain, but the blood vessels, nerves, and membranes are pain-sensitive.

It is estimated that nine out of ten people suffer at least one headache per year, and an estimated 40 million Americans suffer from chronic headaches. The most common of these are tension headaches, which generally localize in the forehead and back of the neck, and feel like a tight band or vise wrapped around your head.

The next most common headache is the migraine, affecting some 25 million Americans, of which 75% are women. Typically, the most frequent type of migraine, called common migraine, feels like a hammer pounding inside your head, often behind one eye. As the headache progresses, it can become a dull generalized ache with a sensitive scalp. Nausea and vomiting may occur. A different form of migraine (classic migraine) can begin first with flashes of light, and progress to transient numbness, weakness, vertigo, or speech difficulties. It is believed that migraines may arise when

nerves surrounding the blood vessels around the brain discharge pain signals, when dilated blood vessels prompt nerve fibers to generate and release unstable levels of certain chemicals, or when certain chemicals (specifically serotonin) become altered.

One of the most painful of headaches is the cluster headache, which usually affects men, and has been described as feeling like a red hot poker stuck through the eye into the head. It can occur in clusters throughout the day or several days, thus its name. There can be symptom-free intervals of weeks to months until the next cluster descends upon the victim. This type of headache is believed to occur due to a malfunction of the hypothalamus.

Epilepsy is another type of brain disorder, this one being characterized by an out-of-control activity of part or all of the brain and, subsequently, the nervous system. It may be compared to a devastating earthquake, except that this earthquake occurs within the body. Epilepsy creates an uncontrolled discharge of brainwave activity, which can be determined with an EEG. There are two basic types of epileptic seizures: generalized seizures, which originate in the brain stem and other deeper structures (such as the thalamus) and, usually, rapidly spread to all parts of the brain at once, and focal seizures, which involve only a portion of the brain.

Multiple sclerosis is another serious disorder that affects the brain, spinal cord, and/or nerves of the body. Symptoms can include abnormal sensations, weakness of arms or legs, difficulty seeing, bladder control problems, and/or difficulty with balance. Multiple sclerosis is characterized by remissions and recurrences over a period of many years, and because of this, in early stages diagnosis may be difficult. In many cases, the initial symptoms improve somewhat or completely, only to be followed at some later time with the same abnormalities or new ones. Although the cause is not known, it is more prevalent among people living in northern Europe and the northern United States. Also, some hypotheses link it to past viral infections or autoimmune diseases; it can possibly be precipitated by factors such as trauma, pregnancy, or infection.

Strokes, the third most common cause of death in the United States, result from a blockage of the blood supply to a certain part of the brain. More than any other organ, the brain depends on a second-to-second blood supply, and it is because of this that strokes occur so quickly. Strokes are a result of a blood clot in an artery (thrombus) such as a brain artery; a blood clot fragment usually from the heart or a carotid artery of the neck (embolus); or the bursting of a blood vessel (hemorrhage). Symptoms can include paralysis of a part of the body, affected speech, affected eyesight, inability to

Inside the Brain and Nervous System

The Brain and Spinal Cord

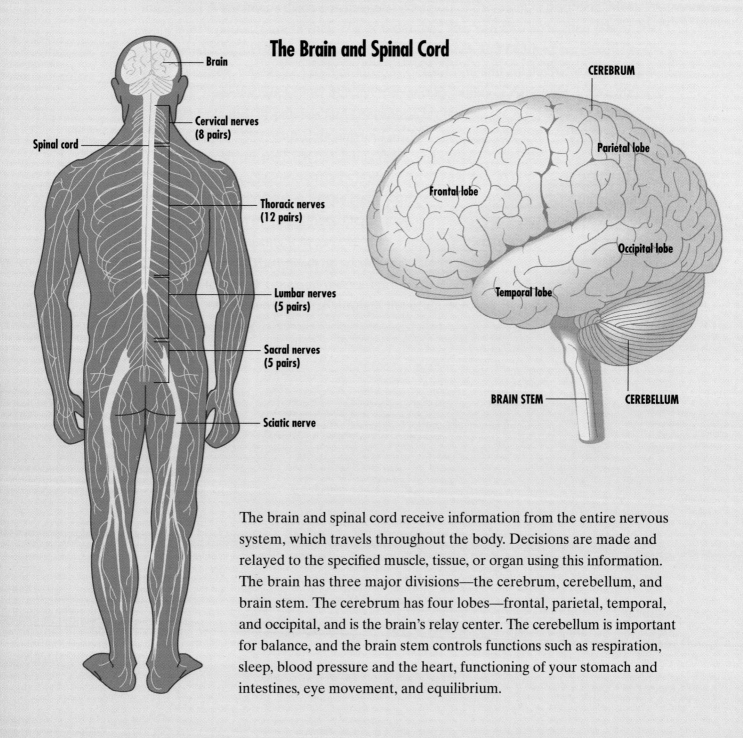

Brain

Cervical nerves
(8 pairs)

Spinal cord

Thoracic nerves
(12 pairs)

Lumbar nerves
(5 pairs)

Sacral nerves
(5 pairs)

Sciatic nerve

CEREBRUM

Parietal lobe

Frontal lobe

Occipital lobe

Temporal lobe

BRAIN STEM

CEREBELLUM

The brain and spinal cord receive information from the entire nervous system, which travels throughout the body. Decisions are made and relayed to the specified muscle, tissue, or organ using this information. The brain has three major divisions—the cerebrum, cerebellum, and brain stem. The cerebrum has four lobes—frontal, parietal, temporal, and occipital, and is the brain's relay center. The cerebellum is important for balance, and the brain stem controls functions such as respiration, sleep, blood pressure and the heart, functioning of your stomach and intestines, eye movement, and equilibrium.

Inner View of the Brain

Cerebral cortex—Important for organization, memory, creativity, and its coverage of the cerebrum. The cerebrum has two hemispheres. The right side of the brain deals with spatial comprehension, musical abilities, and nonverbal ideas; the left side of the brain is important for language and analysis. Each hemisphere controls opposite sides of the body.

Corpus callosum—Acts as a bridge connecting the right and left hemispheres (parts) of the cerebrum.

Thalamus—Functions as a distribution center and transports incoming information from all parts of the nervous system. It is also important in taste, smell, pain, and memory.

Hypothalamus—Functions in temperature, sleep, hunger, thirst, pain, vomiting, and control of the pituitary gland.

Pituitary—Often called the master gland, it is vital in growth, maturation, and reproduction.

Brain stem—Handles basic functions, such as breathing, connects the cerebrum with the spinal cord, and is developmentally the oldest part of the brain. It consists of the midbrain, pons, and medulla oblongata.

Midbrain

Pons

Medulla oblongata

Cerebellum—Involved in balance, and makes sure that movements are carried out correctly.

The Cranial Nerves of the Brain

Smell (olfactory)

Vision (optic)

Eyeball muscles (oculomotor)
Eyeball muscles (trochlear)

Muscles of chewing and biting (trigeminal)

Eyeball muscles (abducens)
Muscles of the face, sensation to the face, and taste buds (facial)
Hearing and balance (acoustic)
Taste and swallowing (glossopharyngeal)
Heart, lungs, and digestive system (vagus)

Neck muscles (accessory)

Muscles of tongue and neck (hypoglossal)

[Continued on next page.]

Inside the Brain and Nervous System

Control Areas of the Cerebral Cortex of the Brain

The cerebral cortex is the largest portion of the brain. Specific areas in the cerebral cortex control a variety of functions of the body, ranging from motor coordination to speech and intelligence.

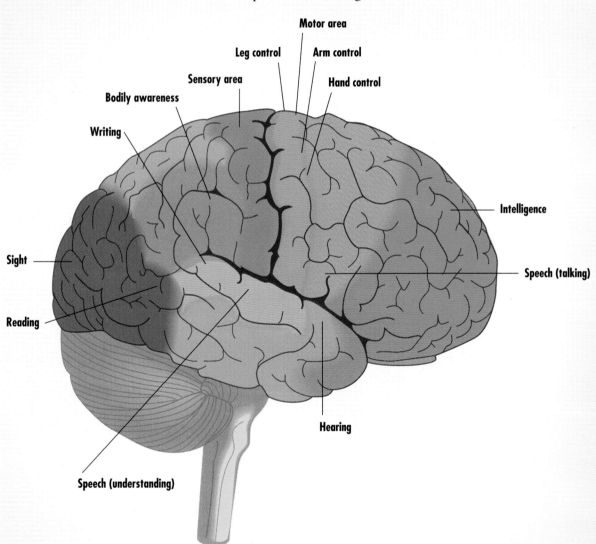

The Sensory Part of the Cerebral Cortex of the Brain

This figure shows in more detail the sensory control areas for different body parts on the sensory area of the cerebral cortex of the brain. Some areas of the body are represented by large areas on the sensory part of the cerebral cortex of the brain, while other parts have a small representation. The lips have the greatest area, followed by the face and thumb, with smaller amounts for the trunk and lower parts of the body. This means that the lips have the largest number of specialized nerves and are the most sensitive.

When Something Goes Wrong with the Brain

Headaches

Tension Headaches These headaches start as a tight band or viselike feeling in the muscles of the back of the head, and spread over the top of the head. The pain can last 30 minutes to 7 days or longer, and the majority of those who have these headaches are women.

Pain often in forehead, temples, or back of head and neck

Pressure on muscle may increase pain.

Trigger factors: muscle spasm, neck arthritis, or jaw/joint pain

Vertigo

Visual auras

Confusion

Poor memory

May have local redness

Disturbed by sounds

Eyes irritated by light

Pale, perspiration

Thick speech

Nausea, vomiting

Migraine Headaches Generally, these headaches produce a pulsating, hammerlike pain, which starts in the temple or the eye and can spread over the whole side of the head. The pain lasts 4 to 25 hours or longer, and again, the majority of those troubled with migraines are women.

Swelling and redness of eyelid

Severe headache behind eye

Pupils constricted and eyes tearing

Flushing of face and sweating

Nose congestion and secretion

Cluster Headaches These headaches appear as a boring, knifelike, excruciating severe pain, usually behind one eye. The pain lasts 30 to 90 minutes or more, and can occur in clusters for weeks to months. Typically, they affect large and strong men.

Strokes

Strokes can be caused by a blood clot in any portion of the brain. This stops the blood supply and creates damage. The damage may be mild or severe, permanent, long-lasting, or temporary. A person with paralysis of the right side of the body has damage on the left side of the brain. A person with a left-sided paralysis has the opposite. In addition to these symptoms, there can be dizziness, unsteadiness, clumsiness of an arm or leg, difficulty swallowing, and thinking or breathing problems.

Damaged area

2 The clot can lodge in an artery in the brain, blocking the blood supply and causing damage to the surrounding area.

Right brain damage **Left brain damage**

Blood clot

1 Different mechanisms can produce strokes. One of them is called an embolus, which begins with a blood clot that has formed in a different part of the body—such as the heart. This clot can then travel through arteries of the chest and neck, and may eventually reach an artery inside the brain.

May have paralysis of the right side

May have difficulty speaking or lose ability to speak

May act slowly and/or cautiously

May have memory deficits in language

May have difficulty performing math calculations or analyzing problems

May have paralysis of the left side

May have spatial deficits (may frequently get lost), and may have difficulty recognizing left side of body

May act quickly and/or impulsively

May have memory deficits in action

May have difficulty with musical abilities or geometric shapes

The Eye

THE EYE, nearly a sphere, may be considered one of the most useful approximate inches (by diameter) on the body. Its main purpose is to allow you to see and perceive the environment around you. This is accomplished in two ways. Objects transmit light to the light sensitive part of the eye, the retina, and then this information is translated by the visual part of the brain.

On the exterior surface, the eye's eyelid protects it and shuts out the environment. Deeper within, each eye has six muscles that rotate the eye horizontally, vertically, and obliquely. When the two eyes are tracking a moving object, the muscles of both eyes act together to keep the object on certain points of their respective retinas. If, for some reason, the coordination between the two eyes is lost, the images no longer fall on corresponding points, and double vision occurs.

Blinking is a reflex to actual or possible insults and also moisturizes the outer surface of the eye. It occurs by contraction of muscles in the upper and lower eyelids. Blinking spreads a protective film of lacrimal (tear) secretions over the surface of the eye and occurs every few seconds. However, reflex blinking can occur as fast as once every 0.1 second.

The body of the eye has three coats. The outer layer is the sclera, or the white of your eye. The front one-sixth of the sclera is called the cornea. Both the cornea and the conjunctiva (inner eyelids) are shielded by a fluid film of tears—secreted by lacrimal glands—and mucous and oily secretions. Tears are essential to wash away irritating particles and fumes. Secretion of tears also occurs with irritation of the cornea, the conjunctiva, or the nose, and with hot peppery foods, vomiting, coughing, and yawning.

The middle layer of the eyeball, the choroid, has an abundance of blood vessels that are important for nourishment to the eye. The front portion of this layer is composed of the iris (the colored portion of your eye) and the ciliary body (which suspends the lens of the eye and contains muscle fibers that help in vision). The degree of pigments determine the color of the iris. The lens, which is located centrally in this area, is composed of a strong elastic capsule filled with transparent fibers. When the lens is in a relaxed state with no tension, it assumes a spherical shape.

Between the cornea and the iris is the anterior, or front, chamber of the eye. This chamber, and the smaller portion behind it, are filled with a clear fluid, the aqueous humor. Behind this front

chamber is the main body of the eye, which contains a transparent jelly—the vitreous humor. The aqueous and vitreous humors are important to maintain a normal intra-ocular (eye) pressure of generally 20 millimeters or less (the figure varies depending upon the instrument used to measure it). In order for this intraocular pressure to re-main normal, the fluid is periodically and automatically drained away. If the pressure becomes abnormally high, then glaucoma, an eye disease, has developed.

The major part of the inner layer of the eye, the retina, is very important because it transmits visual information to the brain and contains cells called cones and rods. Cones control color vision (with photo pigments of red, blue, and yellow) and visual acuity, and rods are important for night vision, light detection, and movement. When a single group of color-receptive cones is missing from the eye, you are unable to dis-tinguish some colors from others. If red cones are missing, you cannot differentiate the spectrum of colors from green through red, with all of these colors appearing as green. If green-sensitive cones are missing, you sense only red in the color range of green through red. If there is a loss of blue receptors, you have a greater quantity of green, yellow, orange, and red colors in your visual spectrum and may not see blue colors well, or at all. (A rapid method for determining color blindness is based on spot charts.) Rods contain only one pigment, called rhodopsin.

The cells of the retina are also important for projecting visual information to the optic nerve of the eye, and then to the brain. Acuity is the eye's response to a visual image and ultimately determines how well you see. Vision charts are used by ophthal-mologists (eye physicians) and optometrists to determine the extent of your visual acuity; 20/20 vision is considered perfect. If you have 20/20 vision, it means that you are able to see clearly at 20 feet what a normal eye can see at 20 feet. Other potential visual abnormalities can involve visual fields, color vision (described above), and depth perception.

A visual field is the area seen by an eye and appears on the retina. The amount of this field is limited by the nose and the eyebrow ridges from above and by the struc-ture of the retina. There are no vision receptors in the area where the optic nerve leaves the retina (to go to the brain). Therefore, if an image falls in this location, it is not seen. This is called the blind spot. Still another component of vision is depth perception, which is the ability of the eyes to perceive distance.

As you age, so does your eye. This can cause significant impairment in some peo-ple. First, there are decreases in visual acuity as a result of changes in the retina, lens, or nerves, and changes in the refractive power of the eye which can lead to nearsightedness

or farsightedness. Also, there often is decreased tear secretion, which produces irritation and discomfort. Although there are some people who continue to enjoy perfect eyesight throughout all or most of their lives, most of us sooner or later develop what is known as nearsightedness or farsightedness.

About 30% of Americans are nearsighted, or myopic. This common disorder prevents one from seeing faraway objects clearly, but the vision of near objects is not affected. These distant objects are blurred, because either the eyeball is too long or the lens of the eye is too curved. Myopia is corrected by the use of glasses or contact lenses. Approximately 60% of Americans are farsighted, or hyperopic. This means that faraway objects are seen well, but close objects are blurred. Farsightedness occurs because the eyeball is too short, or the lens is too weak. It can also be corrected with glasses or contact lenses.

Although there are many eye diseases, we will focus on only a few. The first is glaucoma, which can produce a slow- or fast-rising increase in pressure within the eye. Pressure of the aqueous humor fluid in front of the lens increases pressure in the vitreous humor behind the eye. When the flow of the aqueous humor fluid, within the eye and traveling out of the eye, is decreased, resulting in increased eye pressure, damage can occur to the rods and cones, and there can be destruction of the optic nerve. Blindness can occur quickly if treatment is not provided.

Another common eye condition is the development of cataracts, and it is estimated that 3.6 million Americans have them. They appear as a film on the lens of the eye, which causes a gradual and painless deterioration of sight.

Let us now explore the eye—often called a mirror to your soul—which does act as a mirror to reflect into the body the images of the world around you.

How You See

The eye collects light from objects and projects them on the light-sensitive portion of the eye, the retina. This information is then translated by the visual part of the brain. Color images are formed from both eyes simultaneously, providing a three-dimensional image.

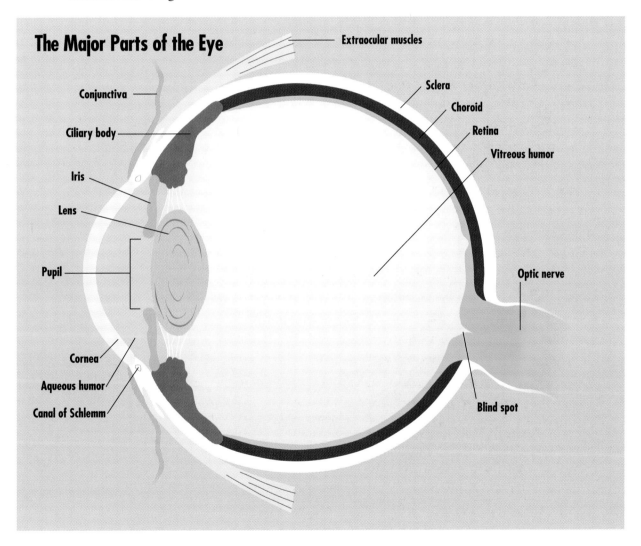

The Major Parts of the Eye

Extraocular muscles

Conjunctiva

Sclera

Choroid

Ciliary body

Retina

Iris

Vitreous humor

Lens

Pupil

Optic nerve

Cornea

Aqueous humor

Canal of Schlemm

Blind spot

As light enters the eye, it passes through the cornea, into the aqueous humor, then to the pupil and lens. The lens of the eye changes its shape in order to focus the light as it further enters the eye. From there, light passes to, and is focused on, the retina; this process is called refraction. Following that, the impulses go to the optic nerve and then to the brain for interpretation.

Looking into Your Eyes with an Ophthalmoscope

Eye doctors look into your eyes with an instrument called an ophthalmoscope. This is what they see—normal arteries and veins, the retina, the optic nerve disc, and the macula, which is said to be the area of the retina with the most "crisp" vision.

Using the ophthalmoscope, doctors can diagnose many eye diseases such as glaucoma, retinal detachments, or hemorrhages. They can also see the effects of systemic diseases that have represented themselves in the eye— diabetes, arteriosclerosis, high blood pressure, syphilis, brain tumors, or brain hemorrhage.

Artery

Optic nerve disc

Vein

Retina

Macula

How the Eye Focuses on Objects

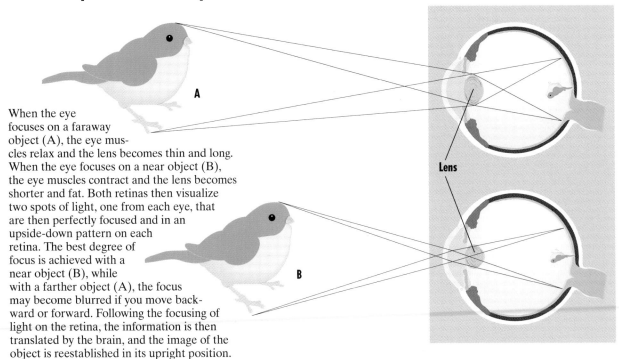

A

B

Lens

When the eye focuses on a faraway object (A), the eye muscles relax and the lens becomes thin and long. When the eye focuses on a near object (B), the eye muscles contract and the lens becomes shorter and fat. Both retinas then visualize two spots of light, one from each eye, that are then perfectly focused and in an upside-down pattern on each retina. The best degree of focus is achieved with a near object (B), while with a farther object (A), the focus may become blurred if you move backward or forward. Following the focusing of light on the retina, the information is then translated by the brain, and the image of the object is reestablished in its upright position.

Nearsightedness and Farsightedness

Nearsightedness (difficulty seeing far objects) and farsightedness (difficulty seeing near objects) can be corrected with eyeglasses. The strength of the proper lens used in eyeglasses is determined by trial and error during an optical examination. First a strong lens is tried, then a weaker lens, and so on, until the lens producing the best vision for you is determined.

Normal eye

Lens

Retina

Nearsightedness (Myopia)

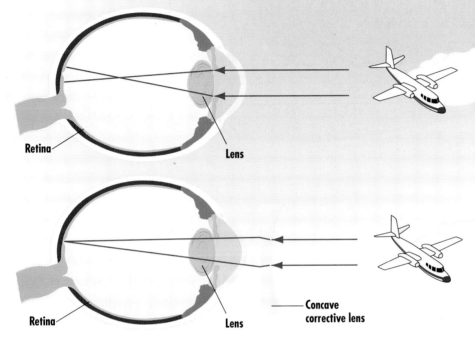

Retina   Lens

The lens in the eye of a person with nearsightedness often has too much refraction, and/or the eyeball itself is too long. This defect produces the image not on the retina itself, but in front of it, resulting in unclear vision.

Retina   Lens   Concave corrective lens

Nearsightedness is corrected with eyeglasses with concave lenses. With corrective lenses, the image is produced on the retina, not in front of it.

Farsightedness (Hypermetropia)

A person with farsightedness has too short an eyeball or too weak a lens. This defect focuses the image beyond the retina, not on it as in the normal eye.

Lens

Retina

Farsightedness is corrected with eyeglasses with convex lenses. With corrective lenses, the image is focused on the retina, not beyond it.

Convex corrective lens

Lens

Retina

Glaucoma

In glaucoma, the aqueous humor fluid has difficulty flowing through the pupil into the front chamber and also in leaving the eye via the canal of Schlemm. This produces excess pressure build up, which can cause damage or blindness.

Primary glaucoma usually occurs without any definite cause. Certain families appear to have a predisposition for this, and it may be more common in people who are far-sighted. Secondary glaucoma is usually a complication of other eye disorders, such as iritis, and people with diabetes or recently controlled high blood pressure can frequently develop this. Glaucoma can occur slowly with minimal or no symptoms, or it can occur within hours, and the person may have severe pain, nausea and/or vomiting, see rainbow-colored halos around lights, and have blurred and/or lost vision. The vision may be permanently lost if proper treatment is delayed.

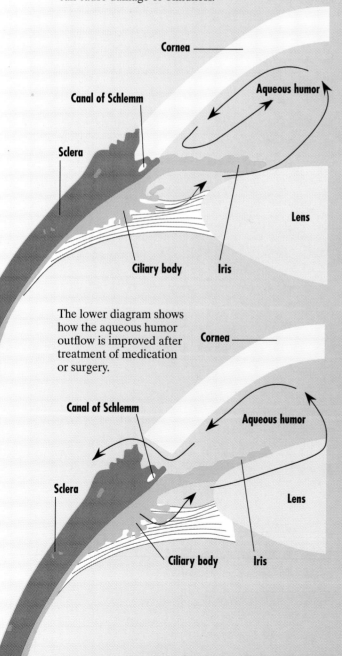

Cornea

Aqueous humor

Canal of Schlemm

Sclera

Lens

Ciliary body   Iris

The lower diagram shows how the aqueous humor outflow is improved after treatment of medication or surgery.

Cornea

Canal of Schlemm

Aqueous humor

Sclera

Lens

Ciliary body   Iris

Eye with glaucoma   **Normal eye**

Glaucoma can cause swelling and stretching of the cornea, as above, so that it can appear very large, and advanced stages can produce blue-gray opacities of the cornea. The pupil of the eye may be enlarged (dilated), will not react to light, and the cornea may be steamy looking with fluid (edema). The diagnosis can be confirmed by using an instrument called a tonometer, or slit lamp instrument, which measures the pressure of the eye.

Cataracts

Cataracts usually develop slowly, without pain, redness, or tearing. They can result from old age, heredity, infection from diseases like German measles early in life, diabetes, eye injuries, or the use of drugs like steroids. The usual symptoms include blurred vision, a feeling of a film over the eye, perceived changes in the color of objects, and/or a need for frequent eyeglass changes.

CENTRALLY LOCATED CATARACT

Retina

Lens

Central cataract area

Aqueous humor

Iris

Retina

Lens

Peripheral cataract area

Aqueous humor

Iris

PERIPHERALLY LOCATED CATARACT

Visually, a cataract appears as a film on the eye's lens.

Cataracts located in the center of the lens can result in more loss of vision than peripheral cataracts.

CHAPTER

3

The Ear

YOUR EAR IS an exquisite instrument with a remarkable design. One of its functions is to hear sounds. It then transmits the sound as impulses to the brain, which interprets and reacts to them. Another important function of your ear is to maintain balance and equilibrium.

Sound is made by nondetectable vibrations of objects (alive and inanimate) that produce a pattern of alternating compression (increased pressure) and decompression (decreased pressure) of the air. The quality of the sound that you hear depends on a combination of frequencies. Music or pleasant sounds consist of one or more related frequencies. Noise, however, consists of a mixture of unrelated frequencies. The intensity of a sound is measured in decibels: the larger the number of decibels a sound produces, the louder the sound is to the ear.

When sound enters your ear and hits the eardrum (tympanic membrane), it puts the eardrum into motion. The eardrum is attached to the smallest bones in your body, the ossicles. They are located in the middle ear and send sounds to the cochlea of the inner ear, the snail-shell-type structure that is an essential organ of hearing. The fluid in the inner ear then causes another membrane, the basilar membrane of the inner ear, to move in a way determined by the frequency of sound. The movement of the basilar membrane stimulates other elements in the inner ear, so that messages are transmitted by the acoustic nerve to the proper area of the brain, the auditory cortex, where the sound is then interpreted.

The ear comprises three main components: the outer ear, middle ear, and inner ear. The outer or *external ear* consists of the funnel-shaped auricle (outside part of the ear), external ear canal (the passageway into your ear), and the cone-shaped tympanic membrane, which prevents objects from entering the ear and keeps the temperature of the eardrum stable. Hairs and wax (cerumen) in the ear canal further help to protect the ear.

The *middle ear*—which has three small ossicles or bones: the malleus, incus, and stapes—is important for allowing movement of the sound messages into the inner ear, to protect the structure of the inner ear from rough movements, and to improve hearing in a noisy environment. The middle ear is also protected by two muscles—the tensor tympani and the stapedius—which act to reduce the transmission of low-frequency sounds. Low-frequency sounds can be excessively loud and may

damage the basilar membrane of the cochlea. These two muscles also function to decrease your sensitivity to the sound of your own speech.

The *inner ear* is essential for both hearing and balance control, and basically consists of the cochlea, semicircular canals, and two chambers—the saccule and utricle. The cochlea, which is responsible for your hearing capabilities, consists of coiled tubes filled with fluid and is separated by membranes. The vestibular apparatus of the inner ear keeps you from being dizzy, and consists of three semicircular canals that govern angular movements of the head, and the saccule and utricle. The horizontal canal governs vertical turning movements, such as dancing the waltz. The vertical canals regulate movements in the horizontal axis, such as leaping and landing after throwing a basketball through a hoop. Balance and hearing messages from the inner ear are then transferred to the brain, via the acoustic cranial nerve, where the brain interprets this information.

Seventy-six million Americans will suffer from an inner-ear disorder, including dizziness or vertigo, at some time in their life. The resulting loss of balance can be frightening: Dizziness is any sense of altered orientation in space (such as lightheadedness, giddiness, or unsteadiness), and vertigo is an illusion of motion (as if the world were revolving around you or you were revolving in space). Other accompanying symptoms for inner-ear disorders can include nausea, ear ringing, hearing loss, weakness, headache, pallor, and perspiration. Inner-ear damage can be caused by infections, injuries to the head, arteriosclerosis, allergies, or neurologic diseases. Tests that may be done to check the amount and type of damage include checking for nystagmus, which are abnormal eye movements, and measuring these eye movements with a test called an ENG (electronystagogram); a hearing test; blood tests; or a CT scan of the brain. Other tests could include platform posturography and/or rotation testing.

Deafness is another ear disease of large proportion. It has been estimated that in the United States alone, there are about twenty-two million people of all ages with significant hearing loss. There are two basic types of hearing difficulties. A problem with the cochlea or the auditory nerve, called *nerve deafness*, can result from the flu, meningitis, other infections, Ménière's disease, arteriosclerosis, certain drugs, or repetitively loud sounds. *Conduction deafness*, an impairment of the middle ear, results from ear wax, infections, genetic problems, or deep-sea diving.

Deafness is usually easy to detect in adults by using standard hearing tests. However, loss of hearing may be more difficult to detect in children. Initially, a child with hearing difficulties may act bored, uninterested, or have difficulty speaking. Often, more sophisticated tests need to be done in young children to measure hearing loss.

We shall now travel through a winding road into the ear and learn more about it.

The Hearing Pathway

Sound waves travel through the external ear, into the middle ear, and then to the inner ear. Once the sound waves reach the inner ear, they are transmitted to the cochlea, which consists of coiled snakelike tubes filled with fluid and separated by membranes. The cochlea contains a structure called the basilar membrane, on top of which lies the organ of Corti. The sound impulses are then transmitted to the basilar membrane, which causes the membrane to vibrate. These vibrations are then relayed to the hair cells within the organ of Corti, which then causes these hair cells to become excited. The total number of hair

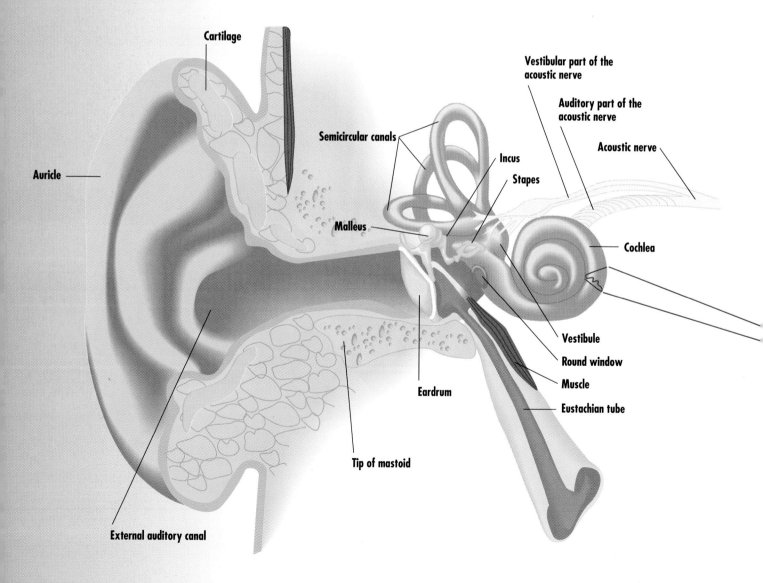

Cartilage

Vestibular part of the acoustic nerve

Auditory part of the acoustic nerve

Acoustic nerve

Semicircular canals

Incus

Stapes

Auricle

Cochlea

Malleus

Vestibule

Round window

Muscle

Eardrum

Eustachian tube

Tip of mastoid

External auditory canal

cells is approximately 20,000, and if damaged, they usually cannot be replaced. Outer hair cells may be important in determining the intensity of sound, and inner hair cells may be important for pitch. The loudness of the tone is also believed to be due to the amount of basilar membrane set into motion and the rate and number of hair cells that vibrate. Impulses from the hair cells are then sent to the acoustic nerve, then to the brain stem, and then to the cortex of the brain. In the brain, these impulses are translated into perceptions and detailed information is presented to you regarding the sound.

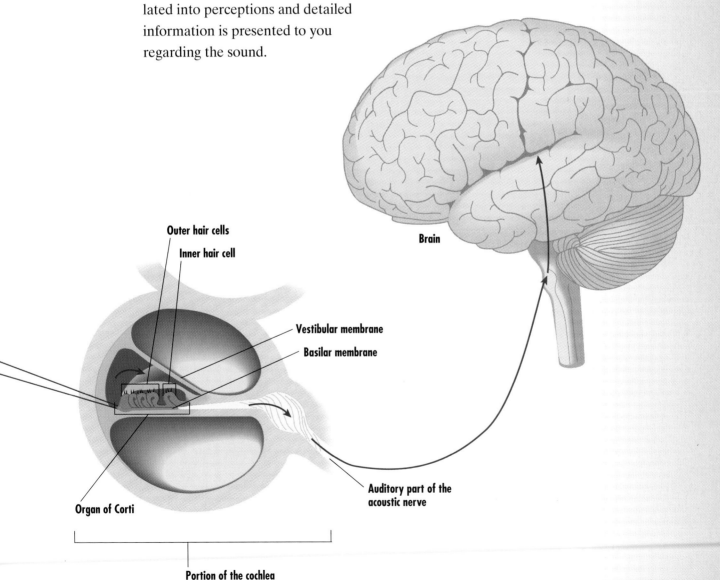

Outer hair cells

Inner hair cell

Brain

Vestibular membrane

Basilar membrane

Organ of Corti

Auditory part of the acoustic nerve

Portion of the cochlea

The Ear and Balance

The main coordinating center for balance is the brain stem of your brain. It processes information from your eyes, neck, muscles, joints, and the vestibular system (balance part) of your inner ear. The inner ear, or labyrinth, is a complicated maze of chambers in the skull. The information from the vestibular system in your inner ear provides critical data on your balance and where you are in relation to the environment.

Brain

Vestibular part of acoustic nerve

Semicircular canals

Anterior

Horizontal

Posterior

Saccule

Portion of a semicircular canal

Crista

Utricle

Spinal cord

The balance part of the inner ear consists of three semicircular canals and two chambers—the utricle and saccule. The semicircular canals are each responsible for a different plane of movement. Each semicircular canal has cristae, or projections, containing hair cells that bend when your head moves. This bending sends nerve impulses to the acoustic nerve, then to the brain stem, and up further into the brain. The utricle is important for proper orientation when you are standing, while the saccule is crucial to your equilibrium when you are lying down. Each chamber of the utricle and saccule has a *macula*, a layer of hair cells coated with a jellylike substance. Every macula also contains otoliths, small granules of calcium, and when the head moves, it also moves the otoliths and hair cells. Again, as a result of stimulation to the hair cells, nerve impulses are sent to the brain along the acoustic nerve. The cerebellum and brain stem are of utmost importance in equilibrium, with the brain providing instructions on how to correct the position of the body, thereby making it more stable.

Diagnosing Balance Disorders

When doctors need to find out the interrelationship between all parts of your balance system—the inner ear, muscles, joints, and eyes—they may use a test called platform posturography for a diagnosis. The patient stands on a computer-controlled floor, and electrodes can also be attached to the person to further monitor their movements. An enclosure with patterns of dots, clouds, or stars surrounds the patient, and a parachute-type harness protects the person from falling when the floor position and the enclosure change. If the vestibular system is not working well, the patient cannot maintain his or her balance. The results can then help the doctors determine the cause and degree of the problem.

Photo courtesy of Equitest® Computerized Dynamic Posturography System, Manufactured by NeuroCom International, Inc., Clackamas, Oregon.

Dizziness and Abnormal Eye Movements

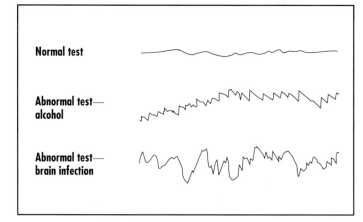

Normal test

Abnormal test— alcohol

Abnormal test— brain infection

Abnormal eye movements (nystagmus) are common in people who have inner-ear disorders, vertigo, or dizziness. These eye movements are involuntary, and alternate between rapid and slow movements. An ENG (electronystagogram) is a common test used for eye movement measurements, and in preparation for this test, electrodes are positioned on the face in specific spots. A technician also inserts water into each ear, at which time the electrodes record the eye movements that may help determine a diagnosis. This illustration shows how the electrodes for the ENG test are often placed on a person's face. The ENG test tracing depicted at the top is normal and is almost flat horizontally. This person had no dizziness or vestibular problem. The middle test result is abnormal and is from a person who drank an excessive amount of alcohol and was dizzy. This tracing shows nystagmus, and each beat of the eye is represented by one upward and downward movement of the tracing. The bottom tracing is also abnormal and is from a patient who had a severe brain infection with damage to the brain and the ear. Nystagmus is also present, but is more disorganized and severe, with no two successive eye movements, or beats, having anything in common with the other.

Your Nose and Senses of Smell and Taste

THE MOST IMPORTANT function of your nose is to breathe in air, and of your tongue is to assist in the swallowing of food; however, their most enjoyable functions are smelling and tasting, respectively. Both the nose and the tongue respond to similiar stimuli, such as the aroma and flavor of a home-cooked meal, and they are often interdependent on each other. For example, the appreciation of that home-cooked meal is enhanced by the aromas permeating from the kitchen, which heightens the taste sensation, and this in turn can intensify the sensation of the aromas coming into the nose.

The purposes of your nose include warming the air, breathing in the air, filtering the air and its particles, and smelling. Anatomically, the nose has two portions, separated by an internal bony partition which is called the nasal septum. The nose also has a number of blood vessels that help to warm the inhaled air. The mucous glands of the nose make sure that the moisture content of the air, which eventually reaches the lungs, is close to 100% and that large particles are trapped and moved out and away, to be coughed up or swallowed. Over 95% of these particles are filtered out, with the hairs at the entrance of the nose removing larger particles, and the adenoids (lymph gland tissue in the nose-throat area) at the exit, filtering smaller particles and organisms. Additionally, your nose defends against infection by moving bacteria and viruses out of the nose, and its mucus also contains antibodies to neutralize these viruses and bacteria. A remarkable one quart of mucus is produced per day!

Another function of the nose, smell, is not as acute as some of our other senses, such as seeing or hearing, and it is poorly developed compared to that of "man's best friend," the dog. In order for you to smell a substance, it must be volatile (so that it can pass through mucus) and it must also be water- and fat-soluble. There are more than 50 different types of olfactory stimulants, although only seven have been defined: camphor, musk, floral, and peppermint, as well as ethereal, pungent, and putrid. The acuity of our sense of smell, just like our hearing and seeing, decreases with age.

Not that much is known about how our sense of smell functions. A current theory is that specified odors attach to nose cells and that causes them to become excited. This excitation causes messages to be relayed to the olfactory nerve and then to the brain. There are extensive connections

between the olfactory (smelling) apparatus and the hypothalamus, brain stem, and other parts of your brain. Doctors and researchers are learning that your sense of smell may be even more important than previously recognized; for instance, certain aromas can bring back past memories of complicated events, which you would not have otherwise remembered. Also, certain smells appear to be involved in the control of your food intake and whether you feel awake, happy, or sleepy. Your sense of smell is most often affected by colds, flu, or allergies. It can also be impaired by head injuries, meningitis or other infections, and Alzheimer's disease.

Another important and enjoyable sense organ in your body is your tongue, which contributes to your sense of taste. Your sense of taste is crucial for the enjoyment of food and drink and to help recognize poisons. Without taste, you could not enjoy an aromatic cup of coffee, a tangy piece of deep-dish pizza, or a sinfully rich hot chocolate sundae. Four basic tastes are determined by your taste buds, which are found on the front, back, and sides of your tongue: sweet, bitter, sour, and salty. Bitterness is the most sensitive taste, followed by sourness, saltiness, and then sweetness. Although all four tastes can be perceived at the tip of the tongue, the tip is most sensitive to sweetness and saltiness. The sides of the tongue are most sensitive to sourness, and the back of the tongue to bitterness. We assume that you can taste more than these four basic sensations, but these are combinations, and involve not only taste, but smell. Taste is just one component of what we call flavor, and flavor involves aromas that travel up the back of the throat to the olfactory (smelling) receptors in your nose. Further research is being conducted in this area.

Taste is determined by taste buds, the specialized nerve endings found on the tongue. You have about 10,000 of these taste buds. Each one is only about 1/30th of a millimeter in size, lives only about ten days, and is then replaced. Unfortunately, the numbers of taste buds also decrease with age. Each taste bud responds primarily to one type of taste, and once a taste is detected, the taste buds send messages to one or more of the cranial nerves, then to the thalamus, and finally to the cortex of the brain for further interpretation.

The most common cause of taste abnormalities is heavy smoking and, in particular, pipe smoking. Impairment can also result from a flu or a cold, dryness of the mouth, deficiencies of vitamins B_{12} or A, cancer, drugs, possibly low zinc, and some unknown reasons.

Now, it is time for us to learn more about how you can smell a sweet red rose, or taste some creamy mint chocolate chip ice cream.

The Olfactory (Smelling) System

Aromas or odors are received by the olfactory (or smelling) nerve cells of the nose. These tapered cells, along with other components, make up filaments that pierce the cribriform plate of the skull—a bony sievelike plate that allows the olfactory nerve filaments to combine and enter the olfactory bulb. From there, a nerve plexus is formed, which sends messages to the brain—the cerebrum, hypothalamus, and brain stem. These olfactory pathways in the brain provide for the control of your food intake, your dislike of the smell of certain foods, and your conscious perception of smells.

Olfactory tract

Area of the cribriform plate

Olfactory bulb

Ethmoid bone

Olfactory epithelium (covering of the inner portion of the nose)

Olfactory nerves

Olfactory areas of the brain

Brain

Frontal sinus

Superior nasal concha

Middle nasal concha

Middle meatus

Inferior nasal concha

Inferior meatus

Opening of nose

Sphenoid sinus

Adenoid tissue

Throat recess

The Sinuses

The sinuses, which lie above your nose, may function in adding resonance to the voice, producing small amounts of mucous, warming and humidifying air, and forming a type of shock absorber. Recent research has shown that even the common cold can produce sinus obstruction in most of us.

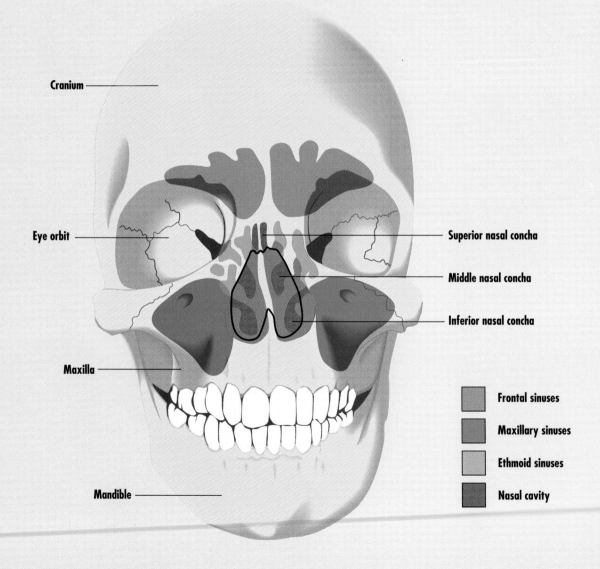

Cranium

Eye orbit

Superior nasal concha

Middle nasal concha

Inferior nasal concha

Maxilla

Mandible

Frontal sinuses

Maxillary sinuses

Ethmoid sinuses

Nasal cavity

How You Know if It's a Pickle or Ice Cream

The Pathway of Taste

Many taste buds are found on the tip, sides, and body of the tongue, although other taste buds, specifically the circumvallate papillae, are located primarily on the back of the tongue, and to a lesser degree on the pharynx (throat), and larynx (voice box). Taste buds are supplied by the facial (entered into by the trigeminal nerve), glossopharyngeal, and vagus cranial nerves, which carry taste messages to the thalamus and then to the cortex of the brain for interpretation. When you eat a pickle, salt taste buds on the tip and sour taste buds on the sides of your tongue are activated and send nerve impulses to your brain. If you eat ice cream, the taste buds for sweetness, which are also found on the tip of your tongue, become excited, and let your brain know how to react to experiencing this taste.

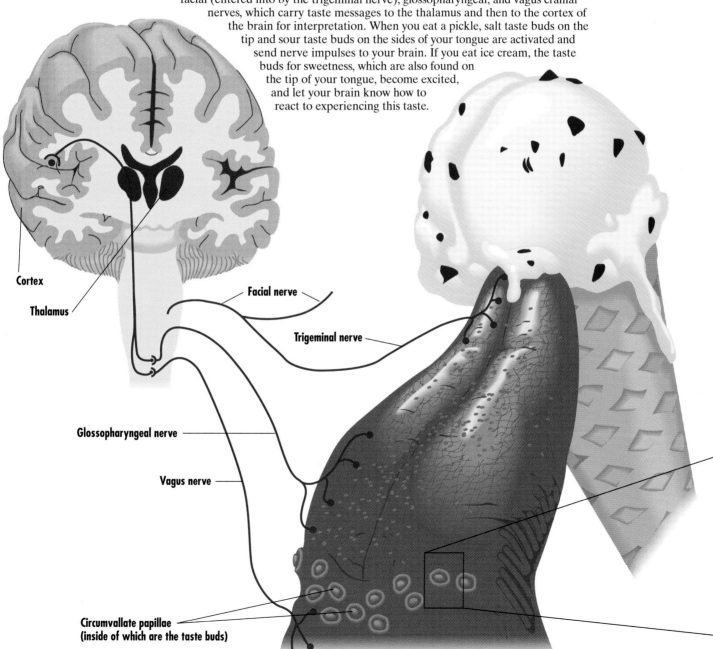

Cortex

Thalamus

Facial nerve

Trigeminal nerve

Glossopharyngeal nerve

Vagus nerve

Circumvallate papillae
(inside of which are the taste buds)

Microscopic Representation of a Taste Bud

Each taste bud comprises approximately 40 taste cells. The outer tips of the taste cells surround a taste pore, and the tip of each cell has protruding taste hairs (microvilli) that point up and out of the tongue. It is believed that these hairs provide the main receptor surface for taste.

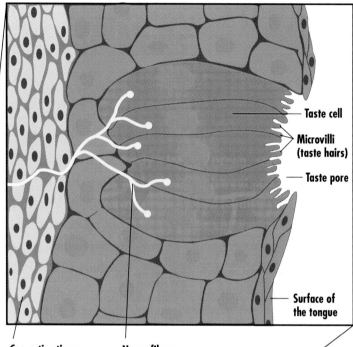

Taste cell

Microvilli (taste hairs)

Taste pore

Surface of the tongue

Connective tissue (denser, stronger cells)

Nerve fibers

Circumvallate papilla

Surface of the tongue

Circular trough

Taste bud

Mucous gland

Lymph nodule

Muscle fibers

Part of the Tongue Showing Taste Buds in a Circular Trough

Taste buds are situated on most of the different types of papillae (projections) found on the tongue. However, they predominantly occur on the walls of the circular troughs of the *circumvallate papillae*, which are mainly located on the back of the tongue.

The Skin, Hair, Nails, and How Your Skin Repairs Itself

The Skin and Hair

This figure shows a magnified version of the skin and its three layers—the epidermis, dermis, and the subcutaneous layer. The hair, glands, and other components are also depicted.

Hair is a derivative of the epidermis of the skin and is made up of the protein keratin. A strand of hair consists of a shaft and a root lying within the epidermis, the hair follicle. The end of the root expands to form a hair bulb. Each hair shaft has three zones—the cuticle, cortex, and the medulla. The cuticle forms the hair surface and consists of keratin. The cortex contains many closely packed, elongated cells, which are also filled with keratin and melanin-containing cells. The central medulla consists of rounded cells containing vacuoles, keratin, and melanosomes (melanin-containing cells). The growth rate of hair varies with the area and thickness of the hair, but generally averages from 1.5 mm to 2.2 mm a week. It also varies with the cycles of growth and hair loss.

The *sebaceous glands* produce an oily substance, sebum, which keeps the skin lubricated.

When *arrector pili muscles* contract, as from cold or emotions, they elevate the epidermis and cause dimpling of the skin, or "goose bumps." These muscles also help to express secretion of the sebaceous glands, which lubricate the hair.

The eccrine *sweat glands* are dispersed in the skin and help to control the body temperature. Another type of sweat gland is the apocrine gland, which is less coiled and less widely distributed. Apocrine glands are usually found in the axilla, genital, perianal, umbilical, and external ear areas.

The *epidermis* is thin and relies on the *dermis* for nutrition. The epidermis is constantly renewed from the bottom up. New cells are formed and push their way up, moving older cells to the surface, where dead cells are sloughed off. This complete process—a cell moving up through the epidermis and being sloughed off—generally takes about 45–75 days, less in areas where the skin is thinner.

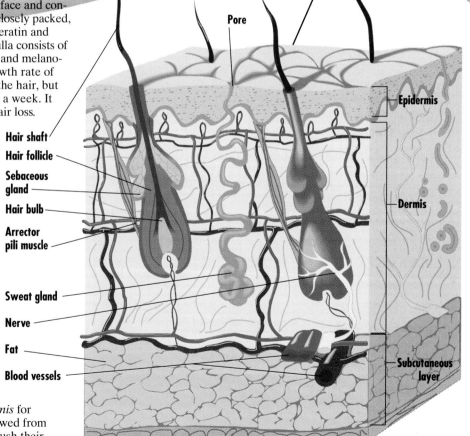

Pore

Hair shaft
Hair follicle
Sebaceous gland
Hair bulb
Arrector pili muscle
Sweat gland
Nerve
Fat
Blood vessels

Epidermis
Dermis
Subcutaneous layer

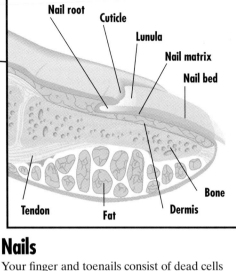

Nail root Cuticle Lunula Nail matrix Nail bed

Tendon Fat Dermis Bone

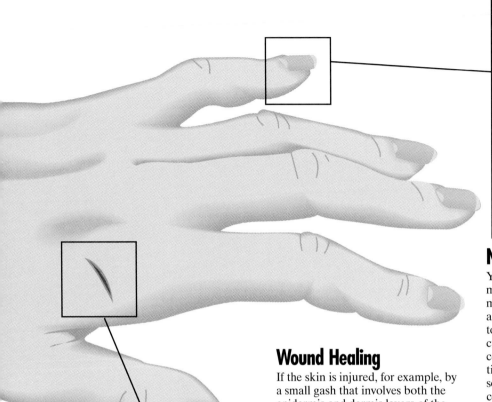

Nails

Your finger and toenails consist of dead cells made up of keratin. The nails contain three major regions—the root, the body of the nail, and the free portion extending over the finger or toe tip. The nail root is inserted into an overlying cuticle. It rests on a plate of living epidermal cells, the matrix, which is joined by the dermis tissue to the finger bone. The matrix extends for some distance beneath the nail's body, where it contributes to the crescentic area, the lunula. The area beneath the nail is rich in blood vessels, and that is why the area directly beneath your visible nail looks pinkish in color. The rate of growth is most rapid in the middle finger—about 0.1 mm per day, and is slowest in the little finger.

Wound Healing

If the skin is injured, for example, by a small gash that involves both the epidermis and dermis layers of the skin, complete healing of the wound usually takes place. The epidermis regenerates cells, while the dermis tries to repair itself by inflammation, reproduction, and remodeling.

Epidermis

Dermis Blood clot White blood cells

Inflammation

1 During inflammation, bleeding in the skin is stopped and is clotted off. White blood cells come to the site to kill off bacteria and to release substances to produce new tissue. Damaged tissues are also taken away.

Reproduction

2 During reproduction, cells and other substances increase, which produces more new tissue and also new blood vessels—capillaries. These substances form a matrix for the second new tissue to fill in the wound.

Remodeling

3 During remodeling, the first new tissue is replaced by scar tissue, with an abundance of collagen, which increases the strength of the wound site.

Moles and Age Spots

Skin moles are areas of the skin where the melanocytes are packed together tightly in a specific area. One of the most common benign skin moles is called seborrheic keratosis and occurs most frequently after middle age, and sometimes develops rapidly in crops. Most moles remain harmless. However, if they are exposed to excessive sunlight, trauma, or abrasion, they may become cancerous. If they become malignant, they can break through the surrounding skin, multiply in other areas of the skin, and possibly travel to other parts of the body and "seed" cancer to other areas. Everyone has skin moles, but when are they cancer? You should consult your doctor if the mole has any of the following characteristics, presented in the format of the American Cancer Society's ABCD's: If the mole is asymmetric, has irregular borders, has color variations (brown, black, white, red, or blue), or the diameter is greater than 6mm (the size of a pencil eraser).

Moles

A mole or a benign (harmless) nevus is a colored, raised area on the skin. Some moles are present from birth, and others develop later. The hallmarks of benign moles are regularity of color, an even surface, and an even border.

Age Spots

Age spots are common, harmless lesions usually found in older people. They are typically found on the back, chest, or face. They appear as plaques that have an almost "stuck-on" appearance, and are brown or sometimes very dark in color.

Skin Cancer

The American Cancer Society recommends these methods to help prevent all skin cancers: Avoid exposure to strong sun, don't use tanning booths, wear clothes to cover you and use a sunscreen lotion when out in the sun, regularly examine your skin for abnormalities, and if you see a suspicious mole, contact your doctor immediately.

Malignant Melanoma

Malignant melanoma is the most important skin cancer because the risk of cancer spread, or metastases, and death is high, unless it is detected and treated at a very early stage. Malignant melanoma can spread to the liver, lungs, bones, eyes, or brain. It has been estimated that by the year 2000, the lifetime risk of developing malignant melanoma may be as high as 1 out of every 75 to 90 people. In men, the cancer typically affects the back and trunk, and in women, the legs, thighs, back, or upper arms.

Basal Cell Cancer

Basal cell cancer is the most common of all skin cancers, with approximately 600,000 cases per year. Fortunately, this cancer has a low potential for spreading to other parts of the body, unlike malignant melanoma. The causes can be many, although it is most often associated with exposure to sunlight, fair complexion, a history of burns, ulceration, or certain scars. Basal cell cancer typically appears on the face or neck and frequently has a pearly border. It slowly increases in size and may become ulcerated.

Squamous Cell Cancer

About 20% of all skin cancers that are not melanomas are squamous cell cancer, and they do have a potential to grow rapidly and to spread to other body parts. Squamous cell cancers also typically occur on the face, especially the lower lip and ears, and often exhibit ulcerations and crusting. The causes include sun exposure, burns, certain viruses, and immunosuppresion.

CHAPTER 6

The Bones

YOUR BONES ARE important because of the structure they provide, and also because of what they contain. They are the framework for the rest of your body. Together they give your body shape, and—moved by your muscles—allow you to run after that tennis ball, take a romantic walk in the woods, or type a letter to your mom. They protect delicate body parts. They also store bone marrow to produce blood cells and contain minerals like calcium and phosphorus that can be released into the blood. It is because of these minerals that bones can survive for many years, even after death.

You have 206 bones in your body, many of which are held in place by strong bands of tissues called ligaments. Both men and women have the same number of bones and the same basic bone structure. However, men usually have broader shoulder bones and a longer rib cage, while women have wider pelvic bones.

The framework formed by your bones is your skeleton, which is divided into three parts. The axial skeleton consists primarily of your vertebral column, skull, ribs, and sternum. The appendicular skeleton consists of the bones of your arms and legs. The ossicle bones in your ears form the third skeletal part (see Chapter 3).

Your bones, as the total skeleton, are able to move because of the subsequent joints that are formed by two bones meeting. Your body parts can have different types of movements depending on the joints involved. For instance, your shoulders and hips have the most freedom of movement because their joints form a ball and socket. Your elbow joint is considered a hinge joint, having a considerable degree of movement but only able to move in one plane or direction. The simplest type of joint movement, usually limited by ligaments or bony processes (protrusions of bone), is a gliding motion typically found in vertebrae and some hand and foot bones.

Now, let us learn more about some of the specific bony areas in your body. The largest bone in your body is the femur or thigh bone, and the smallest bone is the stapes of the middle ear. The skull, or bony part of the head, is divided into two parts—the cranium, which has eight bones and protects the brain, and the skeleton of the face, which consists of 14 facial bones. The vertebral column or backbone comprises the 33 bones called vertebrae. They help support your body, are the basis of your posture, and encase and protect your delicate spinal cord. Individual nerves come out

from the spinal cord, go through openings in the vertebrae, and supply the organs and tissues of your body with valuable information. For example, nerve impulses can be sent to the muscles in your hand to pick up a glass of water, to your legs to run to catch a bus, or to your ribs and diaphragm to assist in breathing. There are seven vertebrae in your neck or cervical area, 12 in your upper-middle back or thoracic area, five in the lower back or lumbar area, five in the pelvic or sacrum area, and four in the tailbone or coccyx area.

Your upper limbs consist of the bones of the arms and hands and the collarbone and shoulder blade, which make up the shoulder. Your lower limbs consist of the hip bones and the bones of the legs and feet. The upper limb bones are designed to give you dexterity, while the pelvis and the leg bones are important for walking and bearing your body's weight.

There are four main types of bones. Short bones are found in the wrist and ankle, flat bones are found in the skull and ribs, long bones are found in the arms and legs, and irregular bones are found in the spine and skull. Most of these bones consist of three basic layers. The first is a thin membrane, called the periosteum, which has nerves and blood vessels. The next is a hard, compact layer called cortical bone. The innermost layer is called cancellous bone. It is spongy, and its inner cavity contains bone marrow, which produces your red and white blood cells.

New bone tissue is constantly being formed and also broken down within your bones. This is achieved by the two basic types of bone cells. Osteoblast cells help to build bone tissue by adding calcium, while osteoclasts break down bone tissue and release calcium into the blood. Your hormones influence this build-up or breakdown of the bones. The female hormone estrogen, the male hormone testosterone, growth hormone from the pituitary gland, and calcitonin from the thyroid glands stimulate the formation of new bone, while the parathyroid hormone from the parathyroid glands increases bone breakdown. Up until the age of about 35, your bones are constantly being remodeled, and the amount of new bone being created is greater than the amount of old bone being broken down. After age 35, your bones have reached their maximum strength and density, and more bone material is lost than is created.

Osteoporosis is a disorder in which calcium and other elements of bone are abnormally depleted, the bone mass is reduced beyond the usual range, and the bone is more brittle. There are two primary forms of osteoporosis. The most common occurs primarily in women after menopause, and the other occurs in men and women over the age of 70. In many cases osteoporosis is a process that begins early in adult life and

progressively worsens; this is especially true in women after menopause. Although osteoporosis can affect men, it primarily affects women over the age of 55. It also targets Caucasians more than African Americans or Asians. The primary reason that women develop osteoporosis after menopause is that the female hormone estrogen, essential for women's bone development, is dramatically decreased after this time. Men have very little estrogen, and the male hormone testosterone, important for their bone development, does not usually decrease precipitously as they age. However, men can develop osteoporosis from conditions related to old age, alcoholism, or other factors. Men or women of any age can also develop osteoporosis from such causes as certain drugs, illnesses, or deficiencies.

People with osteoporosis are predisposed to breaking their bones—most likely, the hips, wrists, arm or leg bones, or the spine. Often, especially if the disease is mild or in an early stage, there may be no symptoms. If the disease is more advanced, the person may become shorter because of collapse of the vertebrae and may suffer from vague backaches, slumped posture, and fractures of the bones. Osteoporosis can usually be diagnosed by the use of x-rays and other tests to measure bone density. Though a tendency to develop osteoporosis may be inherited, in many cases it can be prevented by such measures as adequate calcium intake during one's life, regular exercise, and avoidance of cigarettes and rapid weight loss.

Other common conditions related to the bones are arthritis, scoliosis, neck and backaches, and vertebral bone injuries. There are several kinds of arthritis, of which the two most common are osteoarthritis and rheumatoid arthritis. Osteoarthritis, the most common degenerative joint disease, occurs primarily in older people and most frequently affects the fingers, hips, knees, and spine. It produces pain and stiffness of these joints. Rheumatoid arthritis is an autoimmune disease in which the person's immune system attacks the joints, resulting in inflammation, degeneration, and/or derangement. It can occur in people of any age and typically affects the hands, wrists, knees, and feet. The symptoms typically include pain, stiffness, or limitation of movement of the affected parts. Both types of arthritis usually afflict women two to three times more than men.

Scoliosis is an abnormal curvature of the spine, usually recognized in childhood. It is believed that it may result from abnormalities in the muscles of the spine or in the position of the bones. Most people with scoliosis do not have any symptoms and do not require treatment. However, severe cases can produce back pain, shortness of breath, and occasionally pressure on the lungs or heart.

Neck aches can arise from many situations—serious and not so serious. Usually, they are the result of sleeping in an awkward position or a temporary spasm. They can also result from whiplash or other injuries, arthritis, or a herniated disc. A herniated disc occurs when a disc between your vertebrae ruptures, and the ruptured part slips out of place, causing pressure on a nerve and causing pain. Back pain is most commonly found in the lower region of the spine and is typically due to an injury or stretch of a muscle, ligament, or nerve. It can also be the result of arthritis or herniated discs.

Fractured vertebrae of the spine—from car accidents, sports injuries, or other forms of trauma—are very serious. If this type of injury is suspected, the person should not be moved except by trained medical people. Symptoms can include severe pain, inability to feel sensations, or inability to move.

As you turn the page, you will see just how your bones work—inside and out.

The Skeleton

The Skeleton from the Front
Some of the major bones, as seen from the front of the body, are shown.

Skull
Cranium
Facial bones
Ossicle bones
Maxilla bone
Mandible bone
Cervical vertebra
Collarbone (clavicle)
Shoulder blade (scapula)
Breastbone (sternum)
Xiphoid process
Rib
Lumbar vertebra
Pelvic girdle
Sacrum
Tailbone (coccyx)
Humerus
Elbow joint
Ulna
Radius
Carpal bones
Metacarpal bones
Phalange bones
Femur
Knee joint
Kneecap (patella)
Tibia
Fibula
Tarsal bones
Metatarsal bones
Phalange bones

Axial skeleton
Appendicular skeleton

(A)

(B)

(C)

Ligament

Scapula

Muscle

Tendon

Humerus

Bones, Muscles, and Tendons

The figure to the left depicts the connections that take place in your body, in this case the shoulder joint. As muscles approach bones, the muscle tissue changes into tendons, which attach directly to the bone and provide extra strength, stability, and movability. In some parts of the body, movability is not as important as stability and security, and so ligaments connect one bone to another.

The Joints

The figures to the left show mechanical depictions of three different types of joints.

Figure A depicts the shoulder, a ball-and-socket joint and the most moveable type of joint. The shoulder is formed by the head of the humerus bone fitting into the cup-like joint cavity, thus the name ball-and-socket joint. The shoulder allows you to move your arm toward or away from your body, rotate your arm, or move it in circular motions.

Figure B illustrates the elbow, a hinge type of joint. The elbow allows you to extend, flex, and move your forearm and hand so that your palm faces up or down.

Figure C shows the spine's bones or vertebrae, which are a gliding type of joint. The vertebrae glide on top of each other, making flexion, extension, and rotation possible.

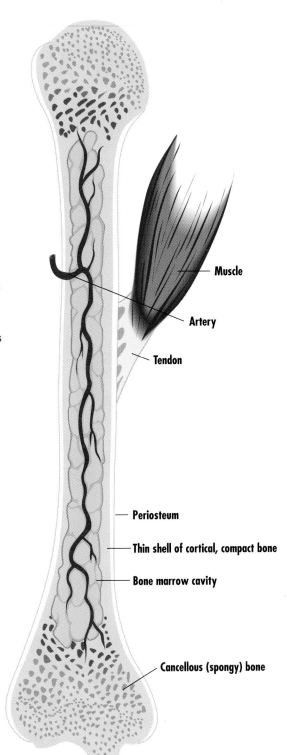

Muscle

Artery

Tendon

Periosteum

Thin shell of cortical, compact bone

Bone marrow cavity

Cancellous (spongy) bone

The Structure of Bone

Bone has three layers—the periosteum, or membrane for nerves and blood vessels; the compact or cortical bone; and the innermost cancellous bone, with the bone marrow cavity where blood cells are formed.

Osteoporosis

Causes of Osteoporosis

The most common cause of osteoporosis is a decrease in the amount of estrogen hormone as a result of menopause in women. Other causes are:

Gland abnormalities (such as estrogen deficiency from menopause, testosterone deficiency, or an increase in thyroid or parathyroid hormones)

Confinement to bed

Poor diet (deficiency in calcium, protein, etc.)

Medicines (such as steroids) or alcohol

Prolonged use of casts (osteoporosis occurs at the site)

Paralysis

Genetic diseases

Space travel

Bone tumors

Chronic disease

The Vertebrae and Osteoporosis

The bones of osteoporotic people are more fragile because less bone mass is present. The vertebrae of the spine lose their height and collapse, sometimes leading to compression of surrounding nerves, fractures of bones, and severe pain.

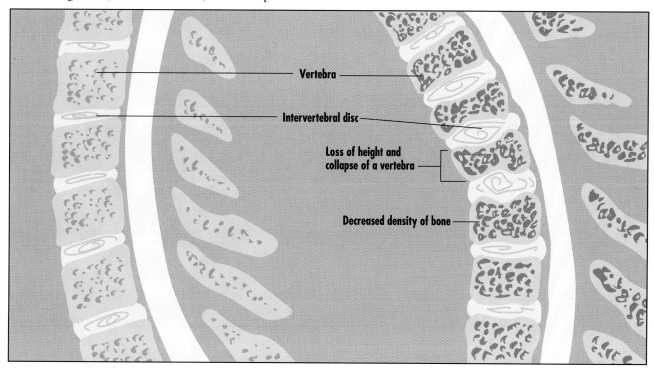

Vertebra

Intervertebral disc

Loss of height and collapse of a vertebra

Decreased density of bone

Normal spine

Osteoporotic spine

Osteoporotic Bone

These illustrations show how osteoporotic bone (right) has a "moth-eaten," more barren appearance compared to normal bone (left).

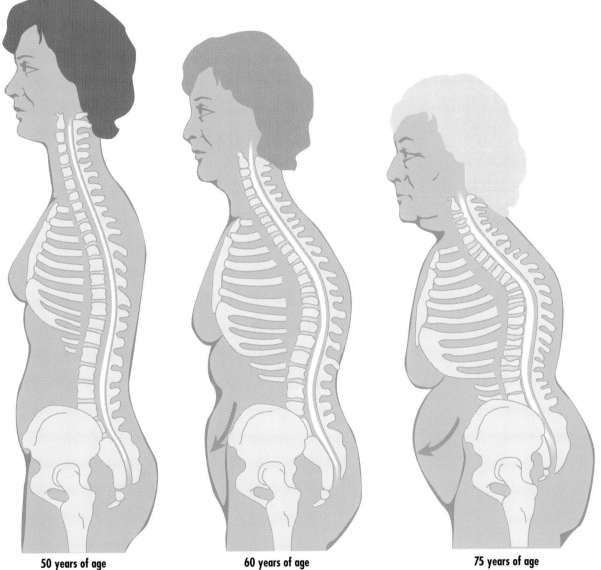

50 years of age   **60 years of age**   **75 years of age**

Osteoporosis and Body Posture

As osteoporosis continues to develop, not only do the vertebrae lose their height, but the person actually becomes shorter. There is a progressive slumping of the shoulders and back, which is called a dowager's hump. If this slumping continues, the lower ribs eventually lie on the hip bone, and the resulting pressure changes cause a ballooning of the abdomen.

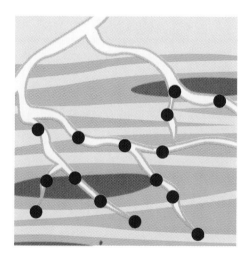

CHAPTER 7

The Muscles

THE MUSCLES IN your body are crucial to various types of movement—voluntary and involuntary. Without muscles, Michael Jordan could not have soared on the basketball court, Fred Astaire could not have danced the waltz, Janet Jackson or Frank Sinatra could not sing, and you could not swallow a delicious meal or laugh at a joke. Muscles help you act and react in your world.

There are three different muscle types—skeletal, smooth, and heart. Skeletal muscles make up about 40% of your body and are generally under your conscious control. They primarily help to regulate movements of the head, neck, back, arms, and legs, and in doing so, they control posture and help you move about. Heart and smooth muscles make up about 10% of your body weight. The heart or cardiac muscles are found only in the heart and have a slower contraction than the other muscles. The heart muscles are controlled by your heart's own pacemaker (see Chapter 10) and are not under your willful control. Smooth muscles also are not under your voluntary control. They line organs and other tissues such as the stomach, intestines, blood vessels, bronchi, gallbladder, urinary bladder, pupil of the eye, and anus. Obviously, smooth muscles are involved in varied processes—digestion, control of blood pressure, and control of urine release, for example.

Though similar in basic ways, the three muscle types do perform differently. Skeletal muscles always have two points of attachment to bones—one called the origin and one called the insertion. This is important for fixation and movement of the muscles. Your muscles are attached to these bones via dense, white, glistening bands called tendons. Nerves are usually embedded within the muscles, and they transmit important information to the muscles about what the body should do. Muscles move thanks to these nerve messages, which can be triggered by many different stimuli. A less intense nerve message produces a weaker muscle response of short duration, and a more intense nerve message initiates a longer and stronger muscle response.

The movement of your skeletal muscles takes place in the following manner. A nerve carries messages from your brain to the muscle fibers, releasing a substance called acetylcholine, which acts on muscle fibers to create what is called an action potential. An action potential is similar to an electrical stimulus and also sends a message—in this case, causing large amounts of sodium ions and then later calcium ions to be released into the muscles. These ions, in turn, produce attractive forces of the actin and myosin within the muscle fiber. This attraction causes the actin and myosin

to slide together, the muscle contracts, and the relevant arm or leg is able to move. Another chemical is later released to permit the actin and myosin to separate, the contraction stops, and the muscle relaxes. When death occurs, all the skeletal muscles of the body contract and become rigid, which is called rigor mortis. This is the result of the body's failure to release the chemical that normally separates the actin and myosin.

The skeletal muscles in your body are constantly remodeled in response to use, disuse, and other factors. When body builders build up specific muscle groups, they are enlarging individual muscle fibers, as well as the total mass of that muscle. This is called muscle hypertrophy. If a muscle is not used over a long period of time, the number of myofibrils (slender muscle threads within a muscle fiber) and the proteins within them decreases, the muscles become weaker and smaller, and then muscle atrophy occurs. If a muscle loses its nerve supply, as from a stroke or injury, it can no longer receive contractile messages, and atrophy can begin immediately. However, if the nerve supply grows back within three months, function can return. After that amount of time, there is less return of function, with generally no return of function after one to two years.

Like skeletal muscles, smooth muscles contain actin and myosin filaments, though smooth muscles differ from skeletal muscles in that smooth muscles have fewer myosin filaments, a slower onset to contraction, as well as a stronger and more prolonged contraction. It is because of these different properties that your stomach muscles can contract to mix, digest, and propel the food to the small intestines, and your blood vessels can be normal or dilated one minute, but very quickly constrict when you leave your warm house and go outside into the cold without a hat or coat on. Smooth muscles also rely on another component, dense bodies, for contraction. The dense bodies are attached to actin filaments and to other proteins that link one dense body to another. It is because of these attachments that the muscle contraction can be spread from one part to the entire muscle or organ. Still another difference is that smooth muscles can be made to contract not only from the action of the nerves but also from other stimuli such as with hormones, decreased body temperature, or changes in the levels of carbon dioxide or other body chemicals. This is important because it allows for rapid changes in the size and activities of the tissues and organs that are made up of smooth muscles.

Heart muscles look and act more like skeletal muscles, and they have similar actin and myosin filaments. However, heart muscles are more interconnected than those of either skeletal or smooth muscles, and because of this, heart muscle contractions can last a long time.

Disorders related to muscles can be divided into those resulting from injury (the most common) and those resulting from disease. A strained or pulled skeletal muscle

(a shoulder muscle, for example) could result from stretching too long or too high when painting your house, a strained calf muscle could result from running a marathon when you're not properly conditioned, or a strained back muscle could occur from lifting a package that is too heavy. Muscle strains may produce ruptures of some fibers within the muscle tissue, which can initiate pain and stiffness. However, this commonly improves itself in a few days or weeks (depending on the severity), as the tissue repairs itself and the swelling and inflammation subsides.

A sprain does not directly involve a muscle, as some people think, but instead is due to the stretching or tearing of ligaments that hold joints together. This can cause swelling, pain, and tenderness in the vicinity, but it slowly resolves as the ligament strands grow back together, and the swelling and inflammation again dissipate. See Chapter 9 for information and an illustration of ankle sprains.

Perhaps the most well-known muscle disease is the Duchenne's type of muscular dystrophy. It is hereditary and typically becomes apparent in boys between the ages of three and six. This type is rapidly progressive, whereas other types may be less severe and progress more slowly. It occurs in about one out of every 3,500 boys. The disease initially produces enlargement of some muscles (especially the calves), but then degeneration occurs producing a decrease in the number of muscle fibers. This reduction of muscle fibers makes climbing stairs, walking, and other movements difficult or impossible. Those muscle fibers that remain become haphazardly organized, with some fibers being smaller and some larger. In the later stages of the disease, only a few scattered muscle fibers remain in a pool of fat cells.

Muscle pain or weakness are common symptoms of many diseases. Types of flu and other infections are the most frequent cause of muscle pain and weakness. However, these symptoms can also occur from the effects of rheumatoid arthritis, thyroid disease, alcohol or heroin use, or metabolic problems caused by changes in the chemicals of your body (for example, calcium, magnesium, sodium, or potassium). In one way or another, these conditions can temporarily or permanently damage the muscle fibers, resulting in abnormal contractions, weakness, or pain.

In cases where no injury has occurred, people who have muscle pain or weakness may be asked by their doctor to have some further tests. These can include blood enzyme tests, to indicate if muscle damage has allowed certain enzymes to leak; a CT (computerized tomography) scan; magnetic resonance imaging; muscle biopsy; or a test called an EMG, which measures the electrical activity of muscles.

Now, please let your finger and arm muscles do the walking and turn the page so that you can see and learn more about your muscles.

The Anatomy of Muscles

Major Muscles of the Front of the Body

Some of the major muscles of the front of the body are shown, with their primary functions.

Sternocleidomastoid—Bends the head to the same side, rotates the head, and raises the chin to the opposite side.

Serratus anterior—Raises the shoulder and is useful in pushing.

Deltoid—Flexes and rotates the arm.

Biceps—Flexes arm and forearm and turns the hand.

Brachioradialis—Flexes the forearm.

Flexor carpi radialis—Flexes and turns the hand.

Pectoralis major—Flexes and rotates the arm and is useful in climbing.

Rectus abdominis—Compresses the abdomen and flexes the spine.

External abdominal oblique—Compresses the abdomen and assists in urination, defecation, vomiting, the birth process, and in letting out a deep breath. It also flexes and rotates the spine, and depresses the ribs.

Tibialis anterior—Flexes and inverts the foot.

Rectus femoris—Flexes the thigh and extends the leg.

Vastus lateralis—Extends the leg.

Sartorius—The longest muscle in the body, flexes and rotates the thigh and leg.

Vastus medialis—Extends the leg.

Gastrocnemius—Flexes the leg and foot.

Tendons leading to origin

Belly of muscle

All muscles have two points of attachment—a point on a bone from which they originate (the origin), and another point on a bone to which they are inserted (the insertion). For stability and strength purposes, both the origin and insertion areas are generally tendons, which are fibrous cords by which a muscle is attached to bone. The belly of the muscle is the main bulk of the muscle itself.

Figure (A) shows a depiction of the biceps muscle of the arm and figure (B) illustrates the gastrocnemius muscle of the leg, both showing the areas of origin and insertion.

Tendon leading to insertion

(A)

Major Muscles of the Back of the Body

Here are some of the major muscles of the back of the body and their primary functions.

Trapezius—Raises, turns, and lowers the shoulder and turns the face to the same or the opposite side.

Teres major—Extends and rotates the arm.

Infraspinatus—Rotates the arm.

Deltoid—Flexes and rotates the arm.

Extensor carpi radialis longus—Extends and turns the hand.

Extensor carpi ulnaris—Extends and turns the hand.

Flexor carpi ulnaris—Flexes and turns the hand.

Triceps—Extends and rotates the arm and extends the forearm.

Latissimus dorsi—Extends and rotates the arm, draws the shoulder down and back, and helps in climbing.

External abdominal oblique

Gluteus medius—Turns, rotates, and can also flex or extend the thigh.

Semitendinosus—Extends the thigh and flexes the leg. After it is flexed, can rotate it.

Gluteus maximus—Extends and rotates the thigh and braces the knees.

Biceps femoris—Extends the thigh and rotates it, and flexes the leg and rotates it.

Semimembranosus—Flexes the leg, and after it is flexed, may rotate it. Also extends the thigh.

Gastrocnemius—Forms the bulk of the calf of the leg and flexes the leg and foot.

Soleus—Flexes the foot.

Area leading to origin

Belly of muscle

Insertion of tendon

(B)

How Muscles Function

All muscles are made up of fibers, thinner than a hair, that can contract and relax—the basic work of movement. Within each muscle fiber are hundreds to thousands of smaller muscle threads called myofibrils, and each myofibril has about 1,500 thick myosin filaments and about 3,000 thinner actin filaments. These muscle components (the actin and myosin) interact to cause muscle contraction and relaxation.

Skeletal muscles are crucial for movements and are found in areas such as the arms, legs, back, and chest and abdominal walls. They are usually under your direct control. Smooth muscles are found in blood vessels and organs and are not consciously controlled by you.

Voluntary (Skeletal) Muscle Movement

Brain

1 A nerve carries a message from your brain to the muscle you want to move.

2 The nerve releases a chemical, acetylcholine, which creates an electrical stimulus in the muscle.

3 Sodium and, later, calcium ions are released into the muscles, enabling the actin and myosin components of the muscle fibers to slide together.

Nerve

4 The muscle moves.

5 Another chemical is released to permit the actin and myosin to separate, and the muscle then relaxes.

COOKIES

Involuntary (Smooth) Muscle Movement

Contraction is much slower in smooth muscles than in skeletal muscles, but smooth muscles can produce prolonged contractions for hours or even days, whereas the contraction in skeletal muscles is much more brief. Smooth muscles do not have the same arrangement of actin or myosin filaments that skeletal muscles do. Instead, the actin filaments of smooth muscles are attached to structures called dense bodies. Some of these dense bodies are linked to other dense bodies by protein filaments that become bridges. It is through these bridge bonds that contraction takes place.

Outer layer of muscle fibers
Inner layer of muscle fibers
Opening

Small intestine

1 When intestinal muscles are stretched by food, nerve fibers can release the chemical acetylcholine, which creates a stimulus in the muscle.

Nerve fiber

Smooth muscle

3 Calcium is removed from the area, and muscle relaxation then occurs.

Calcium

Myosin filaments
Dense bodies
Protein bridges
Actin filaments

2 Calcium can then be released into the smooth muscle, with contraction taking place. Protein bridges and dense bodies are important in this muscle contraction process.

The elbow is a hinge type of joint between the humerus bone and the bones of the lower arm, the radius and the ulna. When the biceps muscle contracts, the elbow, and thus the arm, bends. When the triceps muscle contracts, the elbow straightens. Four types of motion can be achieved by your elbow—extension, flexion, and movement of your forearm and hand so that your palm faces up or down.

The wrist and hand are the most active parts of your upper extremity. The palm side of your hand comes in the most contact with objects, and because of that, its surface has some special features. The skin of the palm side of your hand is thicker and has a greater number of blood vessels than the back of your hand; it contains no hairs, but it does have numerous sweat glands. The large number of hand muscles and the rich supply of nerves to the hands allows them to produce the most intricate movements of your body, which no other animal can achieve. Of all creatures, only you (and other humans) are able to clench, grasp, squeeze, pinch, flex, extend, oppose, and move your fingers in and out.

The wrist is a joint between the radius and ulna bones, and the hand or carpal bones. The wrist is a condyloid type of joint—that is, one oval-shaped bone fits into the elliptical cavity of another bone. The possible movements of your wrist include flexion, extension, and movement toward or away from your body. Within the inner part of your wrist is a cavelike space called the carpal tunnel, which is bounded by four of the carpal bones (some of which help to form the wrist joint). The flexor retinaculum, a horizontal band of strong fibrous tissue, runs between these four bony areas and forms a sheath, or tunnel. The carpal tunnel transports the median nerve and finger muscle tendons from the forearm to the hand, and it is important not only because of these tissues that lie within it, but also because of the problems that can result if the area becomes swollen or inflamed.

Carpal tunnel syndrome is a wrist disorder resulting from the swelling of this flexor retinaculum. This swelling causes compression of the median nerve, resulting in pain, tingling, numbness, and/or weakness in the wrist, hand, and fingers. The problem is more common in women, and this in part may be due to hormonal changes, as carpal tunnel syndrome often occurs during pregnancy, menopause, or before the menstrual period, when additional fluids may collect in the body and its ligaments. This disorder can also be caused by, and worsens with, the excessive use of the wrist, especially in repetitive activities, such as sewing or computer work. Over time, these repetitive actions can produce injury and/or swelling of the flexor retinaculum and nerve, and may thereby cause the syndrome to occur. Carpal tunnel syndrome may also occur as a

result of fluid collection after injury, infections like tuberculosis, vitamin B6 deficiency, or arthritis. Often the symptoms can affect both hands. Symptoms may travel into the palm surface of the thumb, index finger, middle finger, the inner half of the ring finger, and may even spread into the arms. These symptoms usually occur or worsen at night.

In addition to the wrist disorder carpal tunnel syndrome, some other common upper extremity joint problems that involve the shoulder and elbow are dislocations of the shoulder, frozen shoulder, and tennis elbow. Shoulder dislocations or displacements may affect any of the basic subjoints within the shoulder area, and they usually force the corresponding ligaments and/or bones out of proper position. Mild dislocations usually cause only a sprain (or stretching) of the ligaments holding the joint together, and the position of the joint does not move very much out of proper alignment. A more moderate dislocation causes an actual rupture of the ligaments, with the two bony parts moving more out of alignment. In a severe dislocation, all of the related ligaments can be ruptured, and the two bony parts can be seriously displaced.

Shoulder dislocations, especially of the glenohumeral and acromioclavicular joints, occur frequently in athletic pursuits. Glenohumeral dislocations usually occur because of an extension, rotation, or pulling force exerted on the arm, while acromioclavicular dislocations usually result from a fall on the shoulder.

Participating in sports may cause a variety of other shoulder injuries as well. Both baseball and football require complex movements that can maximally stretch the shoulder joint capsule and muscles, sometimes causing tears of the muscles or other ligaments. Golf, which is usually considered a nontraumatic sport, may also result in injuries to the acromioclavicular shoulder joint. On the other hand, the underhand motion of a softball throw or bowling are not as forceful, and thus, injuries resulting from these activities are less common.

Frozen shoulder is another shoulder abnormality. This may result from an injury or a strain and may follow bursitis, tendinitis, or immobilization. Other possible causes include diabetes, heart attack, and some drugs. It produces pain and stiffness and is caused by the inflammation of the joint capsule. Frozen shoulder occurs more frequently in women, especially after the age of 50.

Tennis elbow is an injury that frequently affects middle-aged men. It is a result of a repetitive strain to the area around the ligaments of the outside edge of the elbow. It usually produces a gradual type of elbow pain and is generally found only in one elbow—typically the right. Not only tennis players, but also carpenters, plumbers, gardeners, dentists, and politicians (from excessive handshaking) are among those who often suffer from this injury.

People can be physically challenged because of impairments of any part of the upper extremity—loss of an arm from an injury, severe arthritis limiting movements of the hands, or a stroke affecting the arm(s), for example. Depending on the type and severity of the problems, people can adjust and compensate in various ways. Special equipment like a long-handled shoehorn, a gripping instrument to more easily grasp and open jars, or even a voice-activated computer are available. Additionally, some people learn to develop greater dexterity in their feet or mouth so that they can write or even paint pictures using their toes or lips. This dexterity may occur as the result of increased strength of the muscles in these areas, and/or more or better nerve transmission.

Please turn the page to learn more about the insides of your upper extremities—and some of the most common problems experienced with them.

The Upper Extremity—
Shoulder, Elbow, Wrist, and Hand

The Upper Extremity

These figures show some of the important muscles, arteries, and nerves of the upper extremity.

Deltoid muscle—Important for flexing and rotating the arm inward, and extending and moving the arm away from the body.

Biceps muscle—Flexes the arm and forearm, and turns the palm of the hand to face upward.

Supinator muscle—Turns the palm of the hand to face upward.

Brachioradialis muscle Flexes the forearm.

Flexor carpi radialis muscle Flexes the hand, and moves it away from the body.

Palmaris longus muscle— Flexes the wrist and hand area.

Coracobrachialis muscle—Flexes and moves the arm inward toward your body.

Triceps muscle—Extends the arm and forearm, and can move the arm inward toward the body.

Brachialis muscle—Flexes the forearm.

Bicipital aponeurosis—Tissue formation where the two portions of the biceps muscle come together.

Pronator teres muscle—Turns the palm of the hand so that it faces downward.

Flexor carpi ulnaris muscle tendon—Flexes the hand, and moves it toward the body.

Flexor digitorum superficialis muscle—Flexes the first and second bones of each finger and the hand.

Flexor retinaculum—A strong band of tissue that is important in the carpal tunnel syndrome.

Flexor pollicis longus muscle—Flexes the first and second bone of the thumb, and flexes the first metacarpal bone.

Musculocutaneous nerve

Median nerve

Ulnar nerve

Brachial artery

Radial nerve

Ulnar artery

Radial artery

Superficial arterial arch

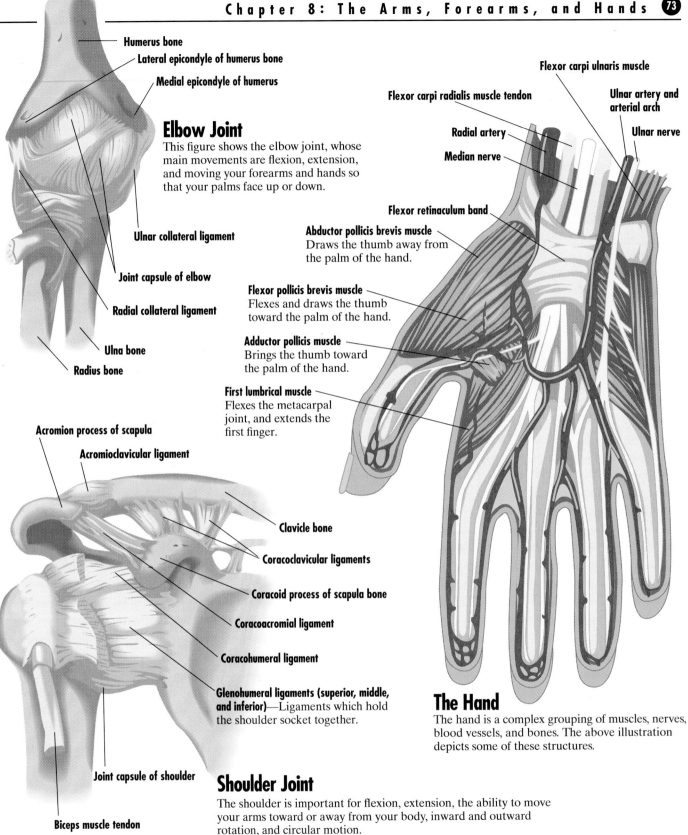

Humerus bone

Lateral epicondyle of humerus bone

Medial epicondyle of humerus

Elbow Joint

This figure shows the elbow joint, whose main movements are flexion, extension, and moving your forearms and hands so that your palms face up or down.

Ulnar collateral ligament

Joint capsule of elbow

Radial collateral ligament

Ulna bone

Radius bone

Flexor carpi ulnaris muscle

Flexor carpi radialis muscle tendon

Ulnar artery and arterial arch

Radial artery

Ulnar nerve

Median nerve

Flexor retinaculum band

Abductor pollicis brevis muscle
Draws the thumb away from the palm of the hand.

Flexor pollicis brevis muscle
Flexes and draws the thumb toward the palm of the hand.

Adductor pollicis muscle
Brings the thumb toward the palm of the hand.

First lumbrical muscle
Flexes the metacarpal joint, and extends the first finger.

Acromion process of scapula

Acromioclavicular ligament

Clavicle bone

Coracoclavicular ligaments

Coracoid process of scapula bone

Coracoacromial ligament

Coracohumeral ligament

Glenohumeral ligaments (superior, middle, and inferior)—Ligaments which hold the shoulder socket together.

The Hand

The hand is a complex grouping of muscles, nerves, blood vessels, and bones. The above illustration depicts some of these structures.

Joint capsule of shoulder

Biceps muscle tendon

Shoulder Joint

The shoulder is important for flexion, extension, the ability to move your arms toward or away from your body, inward and outward rotation, and circular motion.

This figure shows the ligaments of the shoulder. The ligament names indicate where the ligaments come from and go to. For instance, the acromioclavicular ligament means that the ligament goes from the acromion process of the scapula bone to the clavicle bone.

Disorders of the Arms and Hands

Shoulder Dislocation

Shoulder dislocation is a common injury, especially in young men and women who participate in sports. This injury frequently occurs as a result of a fall on the edge of a shoulder or a strong force on the arm.

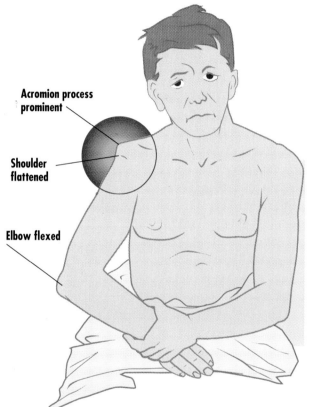

Acromion process prominent

Shoulder flattened

Elbow flexed

Subcoracoid dislocation

This illustration shows a type of common shoulder dislocation, called a glenohumeral dislocation, which involves the shoulder capsule and socket. It depicts a sub-type of this dislocation, which is also common, and is called a subcoracoid dislocation.

Acromioclavicular ligament torn

Coracoclavicular ligaments torn

This figure shows a shoulder dislocation involving an injury to two different types of ligaments. There can be different degrees of severity of all dislocations. For example, a ligament can be stretched or torn, or there may be a more severe injury, as shown here, to the left.

This illustration shows what a shoulder dislocation may look like. The shoulder may be flattened, the humerus arm bone can be prominent, the forearm can be rotated, and the person may use the other hand to support the elbow.

Frozen Shoulder

A person with a frozen shoulder, or capsulitis, has a decreased range of motion of the shoulder. Capsulitis frequently results from injuries, strains, or being immoblized. The joint motions are limited and painful, particularly on rotation. From the rear, there can be a decrease of the muscle mass on the scapula bone and of the deltoid muscles. In the more severe forms, the pain extends into the neck or down the arm.

Here is an inside view of a shoulder affected by capsulitis. It reveals that there are thickenings and inflammation between the capsule, and these adhesions in turn limit the motion of the shoulder.

Adhesions

Carpal Tunnel Syndrome

Carpal tunnel syndrome results from a swelling of the band of tissue lying across the wrist, called the flexor retinaculum. This swelling can then produce compression of the median nerve, which is underneath the flexor retinaculum. At the wrist, the median nerve and muscle tendons pass through a so-called tunnel whose walls are bounded by the carpal bones, and are enclosed by this flexor retinaculum. Carpal tunnel syndrome may involve both wrists and can cause abnormal sensations and weakness in the thumb, index, middle, and the first half of the ring finger.

The symptoms of this syndrome can be reproduced by doctors when they tap on the affected area, or by juxtaposing the hands back to back.

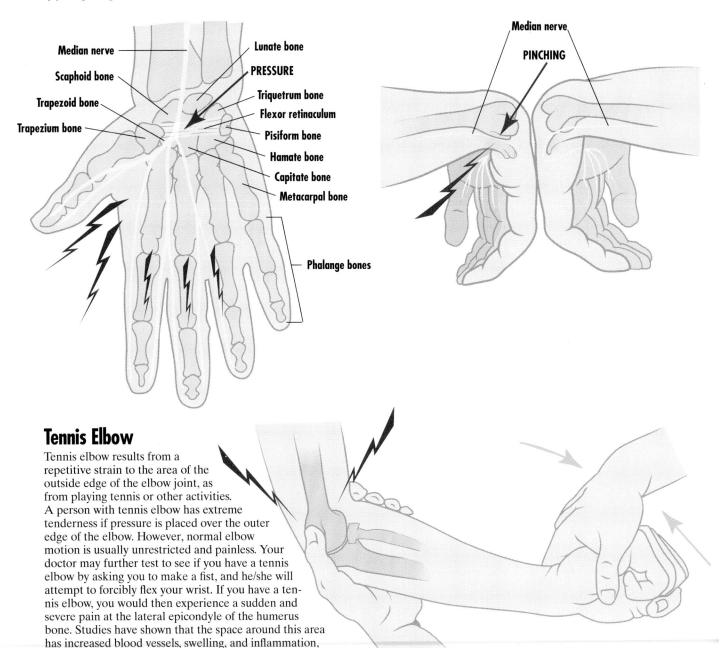

Median nerve
Scaphoid bone
Trapezoid bone
Trapezium bone

Lunate bone
PRESSURE
Triquetrum bone
Flexor retinaculum
Pisiform bone
Hamate bone
Capitate bone
Metacarpal bone

Phalange bones

Median nerve
PINCHING

Tennis Elbow

Tennis elbow results from a repetitive strain to the area of the outside edge of the elbow joint, as from playing tennis or other activities. A person with tennis elbow has extreme tenderness if pressure is placed over the outer edge of the elbow. However, normal elbow motion is usually unrestricted and painless. Your doctor may further test to see if you have a tennis elbow by asking you to make a fist, and he/she will attempt to forcibly flex your wrist. If you have a tennis elbow, you would then experience a sudden and severe pain at the lateral epicondyle of the humerus bone. Studies have shown that the space around this area has increased blood vessels, swelling, and inflammation, but no actual tears of the ligaments or muscles.

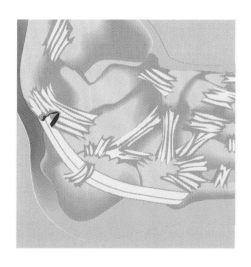

CHAPTER
9

The Legs and Feet

YOUR LOWER EXTREMITY is composed of the thigh, the leg, and the foot. The main joints include the hips, knees, and ankles. Your lower extremity's main functions are to allow weight bearing and walking. This is accomplished by the ankle and the foot, which are the focal points by which the total weight of your body is transmitted into ambulation. Additionally, your foot has a thick heel and toe pads that act as shock absorbers for walking and running. As a child, first you learn to crawl and then to walk. Later, your legs and feet are able to transport you from place to place. It is from this time that you can use your legs and feet to do many useful and pleasurable things—kick a football, ski down a mountain, or run to the lottery office to pick up your million-dollar check.

The bones of your lower extremity include the femur, fibula, tibia, patella, and the foot bones. The femur, or thigh bone, is the longest, largest, and heaviest bone of the body. The tibia, in your leg, is the second largest bone in your body. Also in your leg are the fibula bone, the most slender of all your long bones, and the patella, which is a flat, triangular-shaped bone found in the front of the knee and within the tendon of the quadriceps femoris muscle. The bones of the ankle and foot contain seven tarsal bones (some of which are connected to the tibia and fibula bones), five metatarsal bones, and fourteen phalange bones (two for each big toe and three for each of the other toes).

The hip joint is a ball and socket joint that allows you to flex or extend your thigh, moves your thigh toward or away from your body, and rotates your thigh in or out. The joint is formed by the head of the femur bone fitting into a cup-shaped cavity of the pelvis called the acetabulum. The hip bones consist of the ilium, ischium, and pubis bones. They form a bony ring that supports the vertebrae and rests upon the femur bone of the thigh.

The knee joint is composed of two condylar (knucklelike) joints that occur between the femur and tibia bones and a saddle type joint between the femur and the patella bones. The basic movements include flexion, extension, and slight inward and outward rotation. The knee joint has a number of ligaments that hold it together. Some of the more important ones include the anterior

and posterior cruciate ligaments and the fibular collateral and tibial collateral ligaments. Two cartilages, the lateral and medial menisci (one on the inside and one on the outside of the knee), help to stabilize the knee, allow the joint to slide, roll, and spin, and assist in lubrication and cushioning of the joint. Additionally, the knee has four bursae, saclike cavities filled with fluid that help prevent friction. One is located in the front, one on the inner surface, one on the outer surface, and one in the back. Since the knee (along with the ankle and foot) supports the entire weight of the body, it is particularly vulnerable to injury involving the bursae, ligaments, tendons, and muscles, and other disorders.

The ankle joint is primarily a hinge joint, which permits movements of extension and flexion. However, the ankle joint does not have pure hinge joint qualities, as it also allows slight amounts of gliding, rotation, and movement of the foot toward or away from the body. The hand and foot joints are constructed somewhat similarly. The bones of the ankle or wrist joint consist of cube-shaped bones (carpal or wrist bones and tarsal or ankle bones), which allow a gliding type of movement. The middle group of bones (metacarpal or hand bones, and metatarsal or foot bones) assist the wrist and ankle bones in absorbing the great amount of force put upon them by the body. The toe or finger portions (phalanges) are the most moveable and allow flexion and extension of those parts. However, the hands and feet do differ in that the fingers are much more moveable than the toes, and the bones of the feet are much more solid and less moveable than the bones of the hand.

The great saphenous vein, the largest vein in the body, begins on the inner arch of the foot and goes through the thigh. Within this system of veins and other veins are valves that sometimes become overstretched from increased pressure, as could occur in pregnancy or prolonged standing, which may then result in varicose veins. Varicose veins typically occur in the leg and result from a loss of normal valve activity and abnormal dilation of a vein or veins, especially in the great saphenous vein system (also see Chapter 10). Varicose veins occur more frequently in the legs than in, say, the arms because of the large amount of pressure and weight exerted on the leg valves and veins from the effects of gravity and the distance that the blood must flow from the heart to the leg. The leg veins of a person with varicose veins often appear dark blue and snakelike, and they frequently produce an aching in the leg and swelling of the ankle at the end of the day.

Throughout life there are many conditions that can change the structure and function of the hip and thereby speed up the normal wear and tear on the hip joint. These

conditions include osteoporosis, arthritis, and injuries like hip fractures. A typical person who gets a hip fracture is a woman over the age of 65. The reason may be that women are much more likely to develop osteoporosis, which weakens the bones and increases the chance that a bone will fracture if there is a fall. It is estimated that the hip fracture rate doubles every five years after age 50, and that women have two to four times as many hip fractures as men.

Sprains of the knee ligaments are very common in athletes. The usual cause of a mild to moderate knee sprain is an impact to the back outside portion of the knee while the foot is solidly on the ground. These injuries can result in tenderness, swelling of the injured tissue, a small amount of bleeding in the joint, and a joint that may be more moveable than it should be. Severe sprains, with complete rupture of a ligament, result in the above damage plus an unstable joint. A more severe injury, such as a rupture of an anterior cruciate ligament of the knee, is another that frequently occurs during rough athletic games—when a basketball player twists his extended knee during landing after a jump shot, for example. Often the player hears a pop and feels a strong pain in the knee. Rupture of this anterior cruciate ligament produces a knee that can be very unstable when the person tries to stand on it. The knee "gives way" and may slip or slide, especially when the person tries to turn to the right or left when the foot is still on the ground. This instability is the result of the tibia bone frontally moving out of its normal position on the femur. Another knee ligament, the posterior cruciate ligament, is frequently ruptured during severe extension of the knee or by a blow hitting the front portion of a flexed knee. This type of injury can also occur when you accidentally step in a hole. When the posterior cruciate ligament ruptures, the tibia moves backward out of its normal position, and the front surface of the leg appears to sag. The knee is excessively extended and the joint has more range of movement than it should, causing an unstable knee.

Other knee problems include bursitis, tendinitis, stress fractures, and injuries caused by overuse of the joint. Bursitis of the knee produces an inflammation of a bursa between the patella and the skin. Usually the cause is not known, but it can result from repeated friction to the knee, as can occur in maids or nuns who do repeated kneeling. It can also result from an infection, injury, or diseases such as arthritis or gout.

Tendinitis of the knee causes an inflammation of any tendon in the area. It can result from trauma or overuse of that particular muscle, and may, in some cases, cause calcification and hardening at the knee.

The Lower Extremity, Hip, and Knee

The Lower Extremity

These figures show some of the important muscles, arteries, bones, and nerves of the front and back of the lower extremity.

Sartorius muscle
The longest muscle in the body, and is narrow and ribbon-like. It flexes the thigh, and rotates it inward or outward.

Tensor fascia latae
Flexes the thigh, and rotates the thigh inward.

Rectus femoris muscle
Part of a muscle group that extends the thigh and leg.

Vastus lateralis muscle
Extends the thigh and the leg.

Quadriceps femoris muscle tendon
The tendon of the quadriceps femoris muscle, which extends the thigh and the leg.

Biceps femoris muscle tendon
Flexes the leg, and after it is flexed, it rotates it outward.

Peroneus longus muscle
Flexes the foot downward and lowers the inner part of the foot toward the ground.

Tibialis anterior muscle
Flexes the foot up and inverts it.

Extensor retinaculum

Peroneus tertius muscle tendon
Flexes the foot up, and lowers and twists the inner part of the foot so that it approaches the ground.

Iliopsoas muscle—Flexes the thigh and flexes the lumbar vertebral column to either side.

Adductor longus muscle—Flexes and rotates the thigh, and moves the thigh inward.

Adductor magnus muscle—The upper part rotates the thigh inward, and flexes it. The lower part extends and rotates the thigh outward.

Vastus medialis muscle
Extends the thigh and the leg.

Knee joint capsule

Soleus muscle—Flexes the foot down.

Extensor digitorum longus muscle tendons—Extends the phalanges of the four toes (other than the big toe), flexes the foot up and lowers and twists the inner portion of the foot so that it approaches the ground.

Extensor hallucis longus muscle tendon—Extends the big toe, and flexes the foot up.

Gluteus medius muscle—Draws the thigh out and rotates it inwards.

Gluteus maximus muscle—Extends and rotates the thigh outwards.

Semitendinosus muscle—Flexes the leg and rotates it inward, and extends the thigh.

Semimembranous muscle—Flexes the leg and rotates it inward, and extends the thigh.

Plantaris muscle
Flexes the foot down, and flexes the leg.

Gastrocnemius muscle
Flexes the foot down, and flexes the leg.

Achilles tendon

Femoral artery

Femur bone

Deep femoral artery

Patella bone

Posterior tibial artery

Fibula bone

Tibia bone

Peroneal artery

Tarsal bones

Metatarsal bones

Phalange bones

The Hip Joint

This figure depicts the hip joint. Your hip joint allows you the capability to flex or extend your thigh, move your thigh toward or away from your body, and rotate your thigh in or out.

Anterior superior iliac spine

Ilium bone

Pubis bone

Iliofemoral ligament

Ischium bone

Femur bone

Pubofemoral ligament

Sciatic nerve

Femoral nerve

Lateral femoral cutaneous nerve

Posterior femoral cutaneous nerve

Common peroneal nerve

Saphenous nerve

Tibial nerve

Sural nerve

Medial dorsal cutaneous nerve

Lateral dorsal cutaneous nerve

Lateral plantar nerve

Medial plantar nerve

The Knee Joint

The knee joint, shown here, allows you movements of flexion, extension, and inward and outward rotation.

Femur bone

Posterior cruciate ligament

Anterior cruciate ligament

Lateral condyle of femur bone

Medial condyle of femur bone

Lateral meniscus

Medial meniscus

Fibular collateral ligament

Tibial collateral ligament

Patella bone

Tibia bone

Fibula bone

Patellar ligament

Hip Fracture, Knee Ligament Injuries, and Ankle Sprain

Hip Fracture

Hip fractures can occur in anyone as a result of an injury, but are more common in older women. There are about 250,000 hip fractures per year, and over the past ten years the number of hip fractures has increased. The causes include trauma, injuries, osteoporosis, and old age with a loss of strength in the trabecular bone. However, fractures may also be due to falls resulting from dizziness, vertigo, or stability problems, and from changes in the vestibular system of the ears, heart problems, adverse side effects of medicines, alcoholism, or other health concerns.

This figure shows a fracture of the neck of the femur bone. These types of fractures are usually associated with falls, and are common in the elderly.

— **Fracture of the neck of the femur bone**

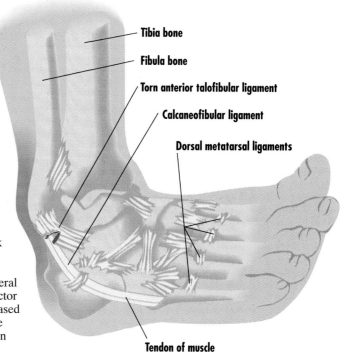

Tibia bone

Fibula bone

Torn anterior talofibular ligament

Calcaneofibular ligament

Dorsal metatarsal ligaments

Tendon of muscle

Ankle Sprain

An ankle sprain can often occur from an injury that involves an inversion of the ankle, with stretching or tearing of a ligament. The anterior talofibular ligament and calcaneofibular ligaments are weak and easily torn, and are usually involved in ankle sprains. A person with a moderate to severe ankle sprain usually has severe pain, swelling from bleeding, and black and blue marks that develop several hours later. To determine if the person has a sprained ankle, the doctor may flex the foot downward toward the ground, and invert it. Increased pain indicates that the ligament may be sprained or torn. The figure here shows some of the ligaments of the ankle and the most common site of ankle sprains, the anterior talofibular ligament.

Femur

Torn posterior cruciate ligament

Torn anterior cruciate ligament

Tibia

Fibula

(A)

Posterior cruciate ligament

Torn meniscus cartilage

Anterior cruciate ligament

(B)

Knee Ligament Injuries

Knee ligament injuries are very common in people who participate in contact sports like football or other sports or activities that place stress on the joints, such as basketball. The injuries can range from minimal stretching to complete tearing of the ligaments. In severe cases, people can have extensive pain, swelling, looseness of the joint, and discoloration of the skin of the knee from internal bleeding in that area. The anterior cruciate ligament (Figure A) is one of the knee ligaments that can be frequently torn during strenuous exercise. If this occurs, the injured person complains that his or her knee gives way and slips or slides with a turn to the right or the left. The posterior cruciate ligament (Figure A) is the main stabilizer of the knee for extension. The most frequent causes of posterior cruciate ligament tears are too much extension of the knee or a blow to the front of a flexed knee. When there is a rupture of this ligament, the tibia bone moves backward, the front surface of the leg seems to sag, and the knee joint is unstable.

Tearing of the meniscal cartilage of the knee is another common injury. The menisci are two c-shaped cartilages, one on the inner part of the joint and one on the outer part of the joint (Figure B). They cushion the knee and are frequently injured as a result of a twisting injury to one or both menisci. People who suffer a meniscal tear develop pain at the knee joint and have an inability to flex or extend the knee properly.

Figure C shows an MRI of a normal knee. You can see the bow-tie shaped appearance of the menisci, which are normal. However, Figure D shows a torn meniscus of the right knee, and here, you can see that the bow-tie shaped appearance is now deranged.

Normal menisci

(C) Normal menisci of the knee
Image courtesy of GE Medical Systems

Torn meniscal cartilage

(D) Torn meniscal cartilage of the right knee
Image courtesy of GE Medical Systems

How the Heart Works

These illustrations show two major, and very connected, activities of the heart: how blood flows through the heart and how the heart beats. The heart is predominantly composed of muscles that act as a pump for circulating blood. Your heart also contains nerves that affect the pumping of the heart by changing the rate of the heartbeat and the strength of the heart's contraction. The heartbeat is a function of nerves and electrical impulses that work together to contract heart muscles and to open and close valves.

The Flow of Blood through the Heart

2 The blood then passes through the right atrium, right ventricle, and pulmonary artery on its way to the lungs.

1 Blood enters the right side of the heart from the superior and inferior vena cava.

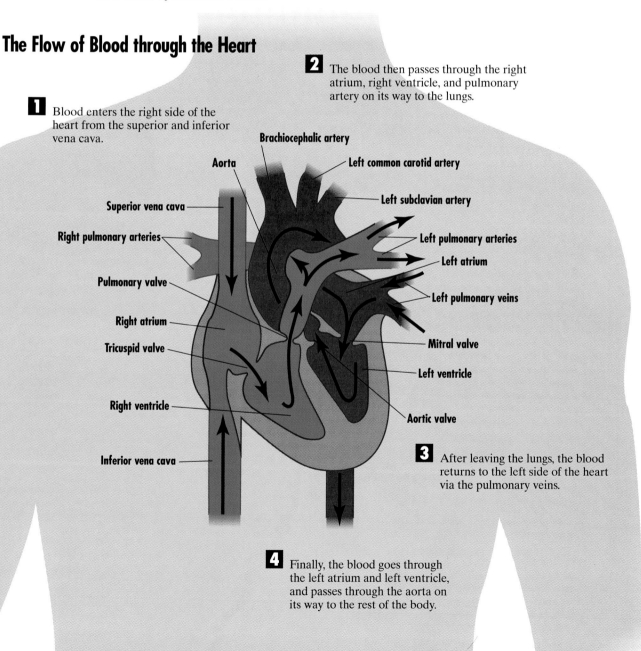

Brachiocephalic artery

Aorta

Left common carotid artery

Superior vena cava

Left subclavian artery

Right pulmonary arteries

Left pulmonary arteries

Left atrium

Pulmonary valve

Left pulmonary veins

Right atrium

Tricuspid valve

Mitral valve

Left ventricle

Right ventricle

Aortic valve

Inferior vena cava

3 After leaving the lungs, the blood returns to the left side of the heart via the pulmonary veins.

4 Finally, the blood goes through the left atrium and left ventricle, and passes through the aorta on its way to the rest of the body.

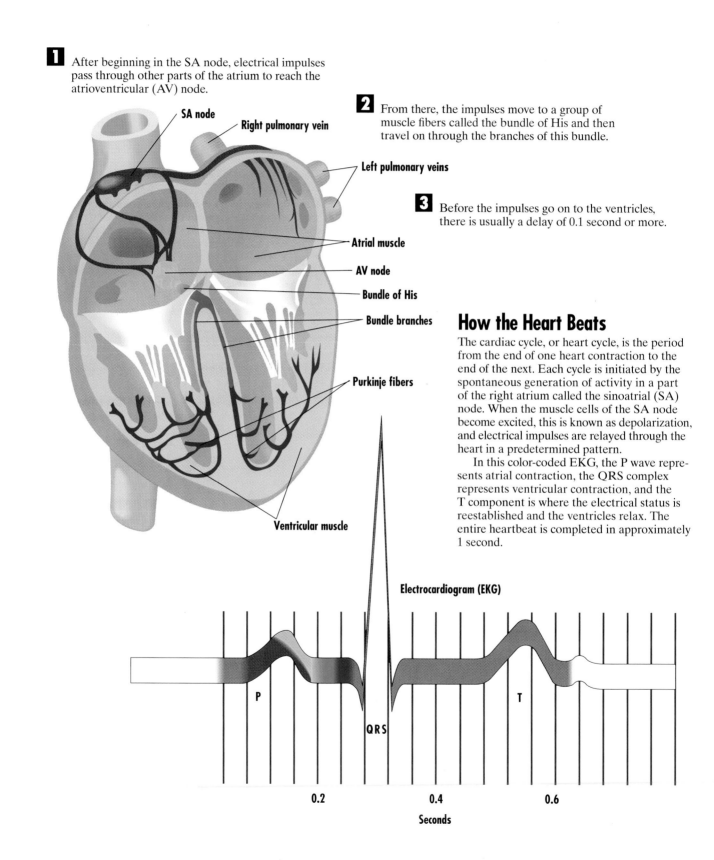

1 After beginning in the SA node, electrical impulses pass through other parts of the atrium to reach the atrioventricular (AV) node.

SA node

Right pulmonary vein

2 From there, the impulses move to a group of muscle fibers called the bundle of His and then travel on through the branches of this bundle.

Left pulmonary veins

3 Before the impulses go on to the ventricles, there is usually a delay of 0.1 second or more.

Atrial muscle

AV node

Bundle of His

Bundle branches

Purkinje fibers

Ventricular muscle

How the Heart Beats

The cardiac cycle, or heart cycle, is the period from the end of one heart contraction to the end of the next. Each cycle is initiated by the spontaneous generation of activity in a part of the right atrium called the sinoatrial (SA) node. When the muscle cells of the SA node become excited, this is known as depolarization, and electrical impulses are relayed through the heart in a predetermined pattern.

In this color-coded EKG, the P wave represents atrial contraction, the QRS complex represents ventricular contraction, and the T component is where the electrical status is reestablished and the ventricles relax. The entire heartbeat is completed in approximately 1 second.

Electrocardiogram (EKG)

P

QRS

T

0.2   0.4   0.6

Seconds

How the Respiratory System Works

Oxygen is critical in providing your body with energy. Although you might be able to survive for weeks without solid food, your body and especially your brain could not last without oxygen for more than a few minutes. The lungs are the organs that facilitate the intake of oxygen and the elimination of carbon dioxide. When oxygen enters the blood in your lungs, it binds with the hemoglobin of the blood and is then transported to your tissues and organs. Carbon dioxide is also brought to the lungs via capillaries in the air sacs, but it is subsequently removed from the body when you exhale.

Respiration involves a cycle of inspiration, in which you breathe air in, and expiration, in which you breathe air out. During inspiration, the rib cage elevates and expands, and the diaphragm contracts. During expiration, the lungs compress, the diaphragm relaxes, and the rib cage lowers and contracts.

The Respiratory System

The diagram below outlines the pathway of a breath of fresh air. The lungs and respiratory system have a number of components and mechanisms for preventing respiratory infections. These include the epiglottis, which is a flap of tissue that covers the larynx and prevents food particles, bacteria, and other infectious organisms from entering the lungs; the production of mucus, which traps the particles and organisms; the white blood cells, which attack organisms that do reach the alveoli; the cilia, which sweep mucus, trapped organisms, and debris toward the pharynx; and the cough reflex, which expels unwanted materials from the bronchioles.

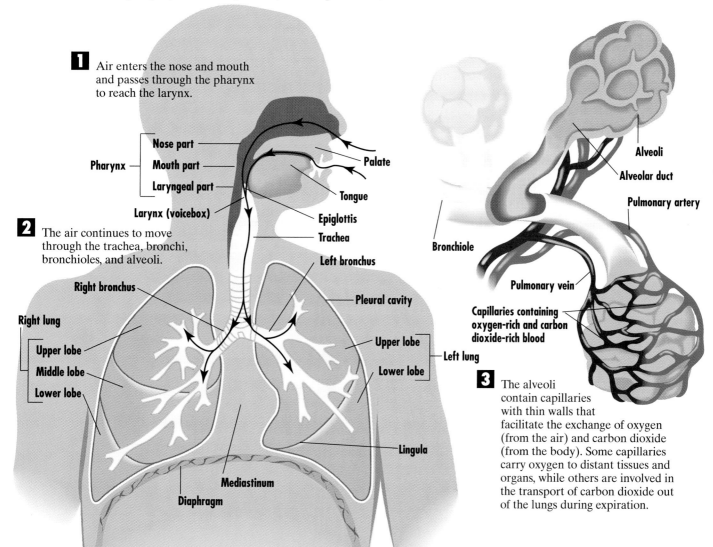

1 Air enters the nose and mouth and passes through the pharynx to reach the larynx.

Pharynx — Nose part
Mouth part
Laryngeal part
Larynx (voicebox)

Palate
Tongue
Epiglottis
Trachea

2 The air continues to move through the trachea, bronchi, bronchioles, and alveoli.

Right bronchus
Left bronchus
Pleural cavity

Right lung
Upper lobe
Middle lobe
Lower lobe

Upper lobe
Lower lobe
Left lung

Lingula

Mediastinum
Diaphragm

Bronchiole

Alveoli
Alveolar duct
Pulmonary artery

Pulmonary vein

Capillaries containing oxygen-rich and carbon dioxide-rich blood

3 The alveoli contain capillaries with thin walls that facilitate the exchange of oxygen (from the air) and carbon dioxide (from the body). Some capillaries carry oxygen to distant tissues and organs, while others are involved in the transport of carbon dioxide out of the lungs during expiration.

A Magnified View of Normal Lung Tissue

This is what a cross-section of healthy lung tissue would look if magnified. The pleura is the membrane that covers the surface of the lung. A respiratory bronchiole is a terminal branch of a bronchiole. Alveoli arise as outpouchings from the bronchioles; clusters of alveoli are called alveolar sacs.

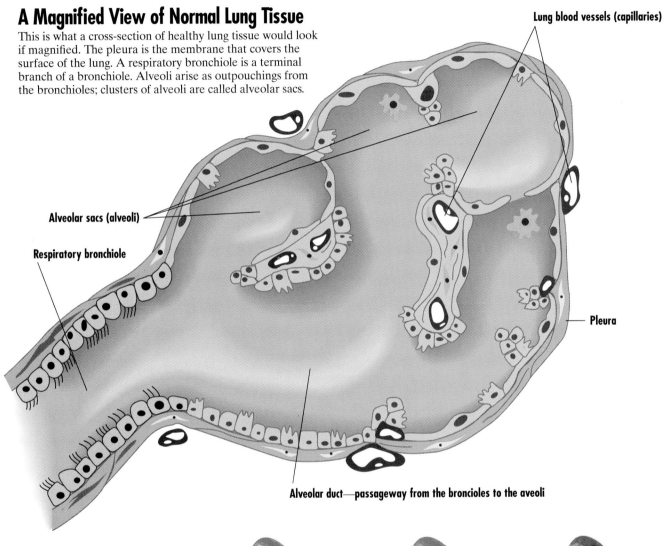

Lung blood vessels (capillaries)

Alveolar sacs (alveoli)

Respiratory bronchiole

Pleura

Alveolar duct—passageway from the broncioles to the aveoli

How Lung Color Can Change with Age

Although lungs are pinkish in color during childhood, they can become grayish-pink later in life as a result of inhaling various environmental pollutants and other contaminants. Each day, your respiratory system is exposed to about 10,000 liters of air that contains chemicals, bacteria, viruses, dust, and other contaminants.

The figures to the right show the possible color changes in the lung from infancy, to middle age, and finally to old age. These changes would vary from person to person and depend on whether the person lived in the city or the country, what their occupation was, and if they were exposed to any other pollutants.

Infant

Middle-aged person

Elderly person

Asthma and Pneumonia

About 5% of all people have asthma at some point in their life. Asthma in children and young adults is generally triggered by plant pollen or dust, while asthma in older people is often provoked by environmental irritants, such as smog. In the early stages of this disease, the small bronchi are plugged with thick mucus. In the later stages, the bronchi and blood vessels are inflamed or thickened, and there can be fluid in the bronchial tissues. Asthma is an episodic disease, with each attack lasting minutes to hours. The disease can be mild or severe, and is sometimes even fatal. During an attack, there is a reduction in the expiratory flow rate and a decrease in the oxygen content of the blood.

Pneumonia can affect any portion of the lungs; it can be mild or severe; and it can be caused by infectious organisms (such as bacteria and viruses), by irritants (such as chemicals), or by food particles or large amounts of saliva that are accidentally aspirated (inhaled) into the lung rather than being prevented (by the epiglottis) from entering the lung. Common symptoms of pneumonia include fever, cough, and chest pain. The diagnosis is usually based on a physical examination and a chest x-ray, although other tests may be required.

Some Things That Can Cause Allergies and Asthma

This tiny organism, the dust mite (as seen under a microscope), is found in common house dust. In some people, it produces an allergic reaction, such as sneezing or an asthma attack.

Pollen grains (shown magnified) are also potential triggers for an asthma attack or other allergic reaction.

People who are allergic to dogs or cats are usually sensitive to the dander (old skin cells that fall off), saliva, or urine of these animals. Immediate hypersensitivity can occur with the symptoms being either mild or severe. Direct contact with the animal is not required to produce the allergic response—only that the person is in an area where the dander, saliva, or urine has collected.

Environmental pollutants, such as concentrated car exhaust and smog from factory exhausts, can increase or precipitate allergic reactions and breathing difficulties. This is particularly true in people with asthma or lung diseases. However, if the pollutant is concentrated enough, it can also affect the general population.

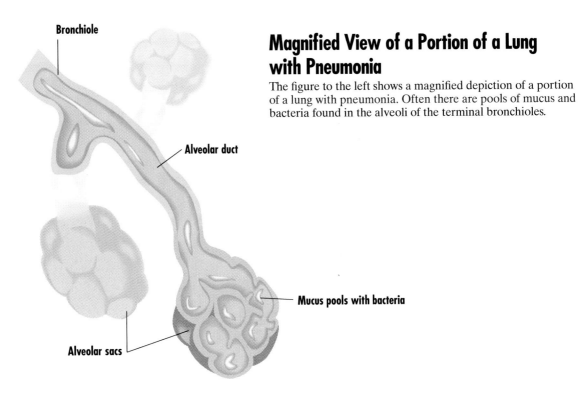

Bronchiole

Alveolar duct

Mucus pools with bacteria

Alveolar sacs

Magnified View of a Portion of a Lung with Pneumonia

The figure to the left shows a magnified depiction of a portion of a lung with pneumonia. Often there are pools of mucus and bacteria found in the alveoli of the terminal bronchioles.

Normal Lung Tissue vs. Lung Tissue with Pneumonia

Both figures below show examples of lung tissue as if viewed under a microscope. The lung tissue on the left contains normal cells (alveoli), while the lung tissue on the right shows abnormalities that are caused by pneumonia. These abnormalities consist of fluid and white blood cells inside the lung cells, as well as edema (fluid) collecting in the lung tissue. All or parts of the lungs may be affected when pneumonia occurs.

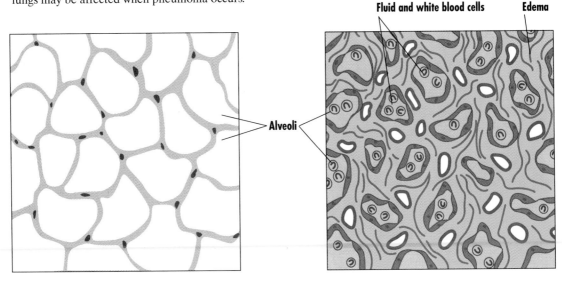

Fluid and white blood cells

Edema

Alveoli

The Lungs and Smoking

An estimated 140,000 people die of lung cancer each year in the United States, and about 85% of these deaths are attributed to smoking. More people die of lung cancer that is related to smoking than of any other type of cancer. Since there has been an increase in the number of women who smoke, the rate of lung cancer in women is increasing faster than it is in men. Lung cancer is now the leading cause of cancer-related deaths in women.

Smoking irritates the bronchi, can constrict the bronchioles, and causes some swelling of tissues. It also causes an increase in the secretion of mucus and other fluids, and since smoking slows down the movement of the cilia, this decreases the ability of the cilia and mucus to remove contaminants.

Emphysema, which is a lung disease, literally means an overabundance of air in the lungs. Its most common cause is cigarette smoking. In the United States, about 70,000 people die each year from emphysema and bronchitis, and an estimated 80% of these deaths are attributed to smoking. Early symptoms of emphysema may include cough, difficulty in breathing, and increased production of mucus and saliva. These symptoms worsen over time, and it becomes progressively more difficult to breathe. The respiratory rate increases, the vital capacity decreases, and the expiratory flow rate decreases.

A Normal Lung and a Lung of a Smoker

Lung tissue is normally elastic and dark pink in color, as shown in the tissue on the left. In contrast, the lung tissue in people who smoke is flabby and blackened, contains carbon particles, and has blocked airways, as shown in the drawing on the right.

Healthy lung

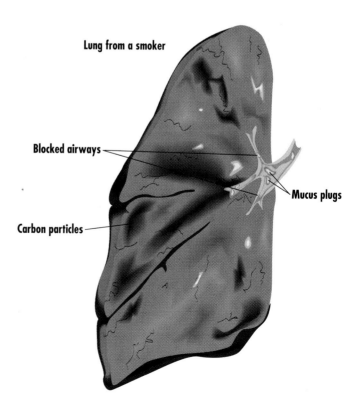

Lung from a smoker

Blocked airways

Mucus plugs

Carbon particles

Image courtesy of Lutheran General Hospital, Park Ridge, IL, and Dr. John P. Anastos, D.O.

CT (Computerized Tomography) Scan of Lung Cancer

This CT scan shows a cancer in the left upper lobe of the lung.

Lung cancer

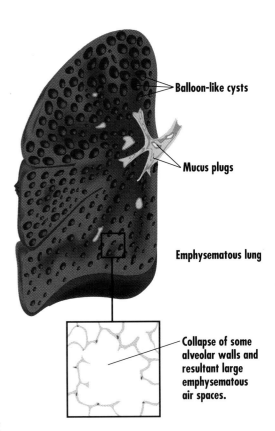

Balloon-like cysts

Mucus plugs

Emphysematous lung

Collapse of some alveolar walls and resultant large emphysematous air spaces.

Healthy lung

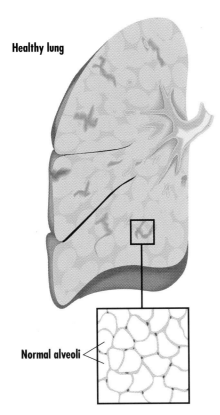

A Normal Lung and a Lung with Emphysema

The drawing on the right shows how tissue from a lung affected by emphysema looks when viewed with the naked eye. The tissue is red-black and contains carbon fragments. Several large cysts, or balloonlike structures, are typically seen and are the result of destruction of the alveolar walls. In comparison, the normal lung tissue shown to the left is pinkish in color, smooth, and has no carbon particles.

Normal alveoli

While normal lung tissue appears compact, as shown in the microscopic view of normal lung tissue on the left, emphysematous lung tissue shows large air spaces, as shown in the example on the right. The over-inflated lung of a person with emphysema has lost its elasticity, the lung has difficulty contracting, air becomes trapped, and further air exchange becomes more difficult.

capacity is only about 30 ml, which increases to about 1500 ml in adults. The functions of the stomach are to temporarily store food, mix and help digest the food with stomach secretions and hydrochloric acid, produce an antibacterial action on its contents, and contract and squeeze the food down to the small intestines.

There is only a small amount of absorption from the stomach into the bloodstream, with most of the absorption occurring instead in the small intestines. However, water, alcohol, and sugar may be directly absorbed from the stomach, and this is one of the reasons that drinking alcohol on an empty stomach can produce an enhanced alcohol effect. Proteins, such as a T-bone steak, are partially broken down by secretions of the stomach. Not much work is required by the stomach for carbohydrate digestion, so this type of food proceeds quickly to the small intestines.

Most digestion takes place in the small intestines. The small intestines are divided into three parts. The first is the duodenum, followed by the jejunum and ileum, with their movements being a combination of mixing and propulsion. Both your small intestines and your mouth digest carbohydrates like rice and pasta. The small intestines also complete digestion of all the fats and proteins.

The appendix is a portion of the intestines where the small and large intestines join together. It is not believed to play a major role in the body, except when it becomes inflamed and appendicitis occurs. In cases of appendicitis, the appendix must be surgically removed to prevent massive body infection.

The next section of the digestive tract consists of the large intestines. They are made up of four sections—the ascending colon, which joins the small intestines, the transverse colon, the descending colon, and the sigmoid colon, which is s-shaped.

All necessary nutrients from the intestines (mainly the small intestines) are absorbed into the bloodstream, metabolized, and then transformed by the liver to be utilized elsewhere or stored there. The liver, the largest organ of the body, is reddish-brown and lies on the upper-right side of the abdomen. It is an important site for the storage of fats, carbohydrates, proteins, and vitamins. The liver is also crucial in that it forms and secretes bile, a mixture of bile acid and cholesterol, which emulsifies fats and neutralizes acid. Additionally, the liver absorbs poisons and toxic substances and neutralizes them.

The gallbladder is a pear-shaped, hollow organ closely attached to the back side of the liver. Its function is to store the bile secreted by the liver.

Some of the most common and talked-about abdominal illnesses, other than minor irritations or flu, include ulcers, gallbladder disease, colon polyps, and cancer.

Indigestion, while one of the most common health problems, is usually just a minor annoyance, and quickly dissipates. It usually occurs from eating too much or too quickly, or eating foods that irritate your stomach. The most common type of stomach disorder that usually does not go away without medical treatment is the ulcer. An ulcer is a scratch or deeper abrasion of the stomach or intestinal lining. There are two basic types of ulcers—stomach and duodenal (afflicting part of the small intestines). Duodenal ulcers are three to four times more common than stomach ulcers, occurring in about 6–15% of the population. However, the incidence of stomach ulcers is generally increasing, while the incidence of duodenal ulcers is decreasing.

Most medical experts now believe that stomach irritation or gastritis, probably ulcers, and some stomach and duodenal cancers may be the result of a bacterial organism called Helicobacter pylori. Also, factors such as genetics, emotions, stress, tobacco, alcohol, and certain drugs such as aspirin or others prescribed for arthritis may cause, precipitate, or aggravate ulcers.

The basic cause of inflammation or ulceration in the stomach or duodenum is not completely understood. However, people with duodenal ulcers are believed to have an imbalance between their protective barriers (of mucus production and secretions of the pancreas and intestines) and their secretion of highly acidic stomach juices. Therefore, these people may produce too much stomach juice or too little mucus. A somewhat controversial view is that stress or anxiety may also precipitate a duodenal ulcer. This may occur as a result of stress signals in the brain stimulating the vagus nerve (a cranial nerve—see Chapter 1), which in turn increases the acidic stomach secretions, which then empty into the duodenum and may produce an ulcer there.

In contrast, stomach ulcers usually do not occur in people with excess stomach secretions, but instead occur from the irritating effects of drugs, alcohol, or cigarettes or from a reduced resistance of the stomach lining to digestion.

Gallstones are a stonelike precipitation that may form in the gallbladder or in one of its passageways. These stones are usually very hard and can obstruct one of the passageways or the gallbladder itself. Approximately 20 million Americans develop gallstones, with about 35% of women and 20% of men having them. Additionally, gallstones have an especially high occurrence rate in Native American Indian women, of whom about 70% develop them. If the stone or stones remain in the gallbladder itself, then there may be few if any symptoms or problems. However, symptoms and problems usually do occur when the stones move and obstruct one of the ducts or passageways of the gallbladder. The symptoms can then include nausea, vomiting, pain, and

The Digestive System and Related Organs

The Digestive System, Liver, and Gallbladder

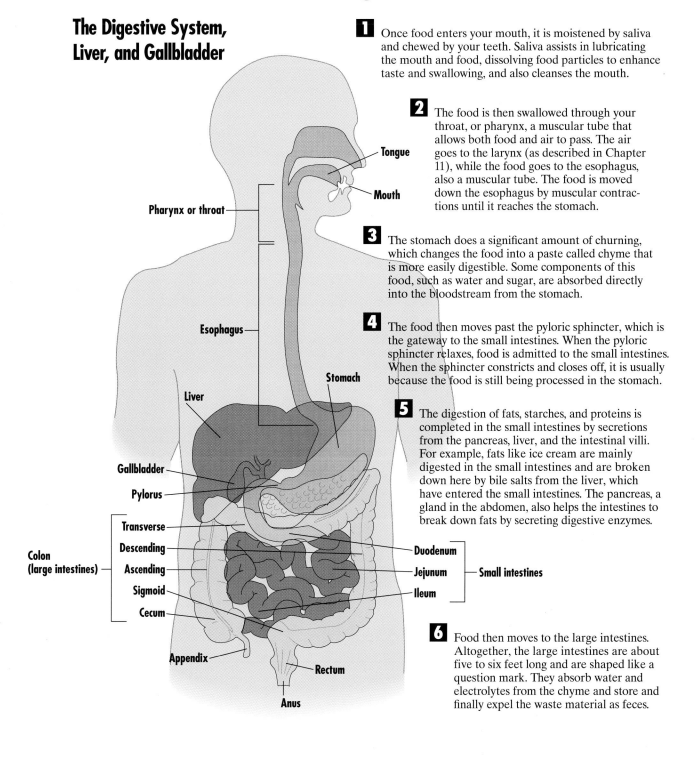

1 Once food enters your mouth, it is moistened by saliva and chewed by your teeth. Saliva assists in lubricating the mouth and food, dissolving food particles to enhance taste and swallowing, and also cleanses the mouth.

2 The food is then swallowed through your throat, or pharynx, a muscular tube that allows both food and air to pass. The air goes to the larynx (as described in Chapter 11), while the food goes to the esophagus, also a muscular tube. The food is moved down the esophagus by muscular contractions until it reaches the stomach.

3 The stomach does a significant amount of churning, which changes the food into a paste called chyme that is more easily digestible. Some components of this food, such as water and sugar, are absorbed directly into the bloodstream from the stomach.

4 The food then moves past the pyloric sphincter, which is the gateway to the small intestines. When the pyloric sphincter relaxes, food is admitted to the small intestines. When the sphincter constricts and closes off, it is usually because the food is still being processed in the stomach.

5 The digestion of fats, starches, and proteins is completed in the small intestines by secretions from the pancreas, liver, and the intestinal villi. For example, fats like ice cream are mainly digested in the small intestines and are broken down here by bile salts from the liver, which have entered the small intestines. The pancreas, a gland in the abdomen, also helps the intestines to break down fats by secreting digestive enzymes.

6 Food then moves to the large intestines. Altogether, the large intestines are about five to six feet long and are shaped like a question mark. They absorb water and electrolytes from the chyme and store and finally expel the waste material as feces.

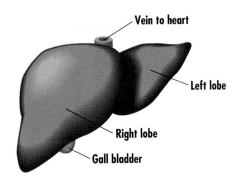

Liver

The liver is a reddish-brown organ, and it transforms and stores nutrients that have entered the bloodstream as part of the digestive process. The nutrients that the liver stores include fats, carbohydrates, proteins, and vitamins; it forms and secretes bile; and takes up and neutralizes toxins.

Stomach

The stomach consists of three layers of muscle covered by peritoneum, the outer membrane that surrounds abdominal organs like the stomach and lines the abdominal cavity. The inner surface of the stomach contains gastric pits that are composed of glands that secrete juices to help in digestion, and it also contains mucus, which protects the stomach lining.

Gallbladder

The gallbladder is a sac that lies beneath the liver and functions to store bile. Bile travels from the gallbladder to the cystic duct and then to the common bile duct, where it enters the duodenum of the small intestines to help in digesting fats.

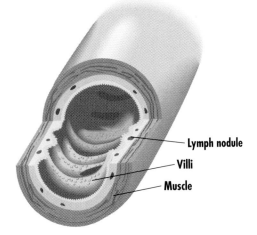

Small Intestines

The small intestines are about 22 feet long. They have circular folds, and their entire surfaces are covered by villi. These villi increase the amount of secretion and absorption that the small intestines can perform.

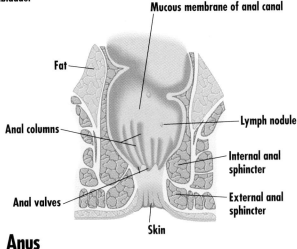

Anus

The anus is the opening of the lower part of the digestive tract from which waste material leaves the body. It is about 1½-inches in length. Hemorrhoids are a common disorder of this area and are due to bulging varicose veins on the inside or outside of the anus.

Gallstones, Stomach Ulcers, Colon Polyps, and Cancer

Gallbladder

Gallstone obstructing cystic duct

Gallstones

Magnification of a gallstone

Gallstones

Gallstones are the most common disorder affecting the gallbladder. Another common disorder is cholecystitis or inflammation of the organ. If a gallstone obstructs the common bile duct, for example, then inflammation of the gallbladder and even the liver may occur. Some 80% of all gallstones are what is known as the mixed type containing cholesterol, calcium, carbonate, and bilirubin.

Stomach Ulcer

Stomach ulcers, also known as gastric ulcers, are usually the result of either an injury (as from aspirin or arthritis drugs), or a defect in the stomach lining's resistance. The types of drugs prescribed for arthritis produce a large proportion of these kinds of ulcers because they reduce the stomach lining's defenses. In fact, stomach ulcers are increasing in frequency, while the other type of ulcers—duodenal or small intestinal ulcers—are decreasing in frequency, possibly due to an increase in the number of people taking arthritis drugs and aspirin, which can irritate or produce ulcers in the stomach.

Most people with stomach ulcers have pain, heart-burn, indigestion, nausea, and/or a loss of appetite. The pain accompanying these types of ulcers may frequently increase after meals because the food stretches the organ, which then increases pain reception. Bleeding is also more frequent with stomach ulcers, and the blood travels down the intestines, where it can then show up as blood in the stool.

Emptying of the stomach contents can be delayed with stomach ulcers. This delay can cause regurgitation of the contents of the intestines and can produce injury and ulceration of the stomach tissues.

A stomach or duodenal ulcer can be diagnosed by a barium x-ray of the stomach or duodenum, or an endoscopy test, whereby a doctor looks into the stomach or duodenum with a lighted tube.

Stomach

Gastric ulcer

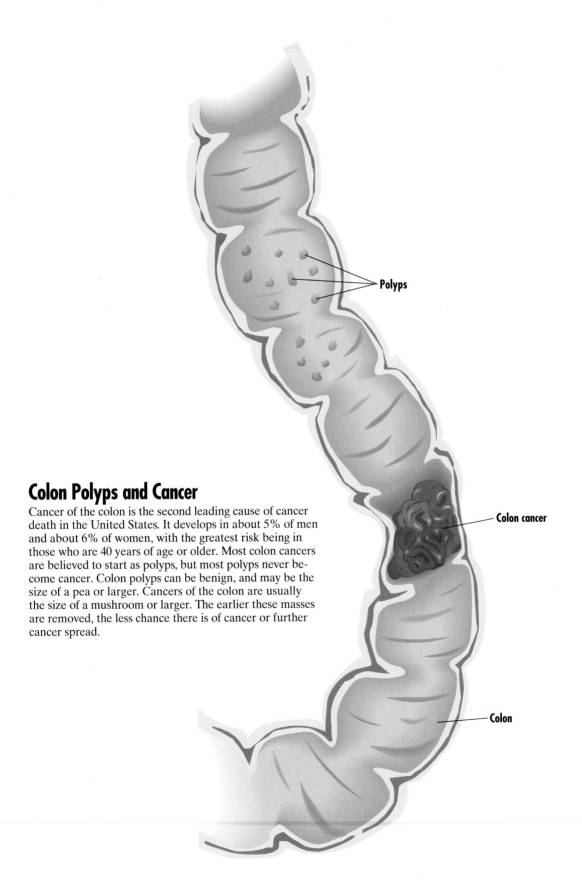

Polyps

Colon cancer

Colon

Colon Polyps and Cancer

Cancer of the colon is the second leading cause of cancer death in the United States. It develops in about 5% of men and about 6% of women, with the greatest risk being in those who are 40 years of age or older. Most colon cancers are believed to start as polyps, but most polyps never become cancer. Colon polyps can be benign, and may be the size of a pea or larger. Cancers of the colon are usually the size of a mushroom or larger. The earlier these masses are removed, the less chance there is of cancer or further cancer spread.

Alcohol and Its Possible Effects on the Body

Alcoholism is a chronic and potentially fatal disease. About 90% of Americans drink alcohol, with an estimated 40–50% of men having at least temporary alcohol-induced problems. Furthermore, 10% of men and 3–5% of women develop alcoholism.

Alcoholism can result in many health problems, which may occur even in the early stages of alcoholism. The severity and number of different symptoms and diseases increase with the length of time a person has been excessively drinking alcohol. The list below shows some of the potential health problems that can result from excessive alcohol drinking. Other problems can include anemia, decreased white blood cell count, heart rhythm irregularities, decreased size of the ovaries, infertility, spontaneous abortion, and fetal alcohol syndrome. Cancer is the second leading cause of death in alcoholics. Their rate of cancer is ten times higher than that expected in the general population, with the liver, pancreas, esophagus, stomach, or head and neck most often affected.

Potential Problems Resulting from Excessive Alcohol Consumption

Hallucinations, paranoia, memory loss, strokes, or brain damage

Yellow sclera of eyes

Painful, swollen muscles

Inflammation of the esophagus with bleeding

Enlarged spleen

Stomach inflammation and bleeding

Pancreas inflammation

Protuberant abdomen

Abnormal liver (cirrhosis, fatty liver, cancer)

Numbness, tingling of the legs and arms

Jaundice or yellow skin

Swollen legs

Alcohol and Liver Disease

Alcoholic liver disease is also a major health problem, with alcoholic cirrhosis a leading cause of illness and death in the United States. The liver's earliest response to alcohol is the accumulation of fat in the liver cells, followed by alcoholic hepatitis, and then liver cirrhosis. A larger number—90–100%—of heavy drinkers have fatty livers, with 10–35% developing alcoholic hepatitis, and 8–20% developing chronic cirrhosis of the liver. Hepatitis or cirrhosis symptoms can include liver enlargement, tenderness, pain, loss of appetite, nausea, and fluid (or ascites) in the abdomen producing a protuberance. These diseases can also produce enlargement of the spleen and inflammation and bleeding in the esophagus and stomach.

The prognosis for a person with alcoholic liver disease improves if the drinking of alcohol is curtailed, before irreparable damage occurs.

Normal liver

Fatty liver

A normal liver is reddish-brown, firm, and nongreasy. During the accumulation of fat in the liver cells from excessive alcohol intake, the liver increases in size and becomes soft, greasy, and yellow.

The word *cirrhosis* has its origins from the greek word "kirrhos" meaning "of yellow color." Cirrhosis is a diffuse, progressive process of the liver characterized by destruction of liver cells by fibrous tissue, followed by scarring and conversion of the normal liver tissue into abnormal hard nodules. The scar tissue impedes the flow of blood through the liver, increasing the blood pressure in the abdomen. This can then lead to multiple other problems including collection of fluid in the abdomen, enlargement of the spleen, dilation of the veins in the esophagus, swelling of the ankles, and kidney problems.

To the eye, the cirrhotic liver looks very nodular, fatty, yellow-orange, and scarred. Alternatively, the liver may instead be small, shrunken, and hard, or large and fatty.

Cirrhotic liver

Image courtesy of GE Medical Systems

R   L

This figure shows a CT scan of a liver cancer. Liver cancer can develop from alcoholic liver cirrhosis

— **Liver cancer**

The Kidneys and the Urinary Tract

1 Urine is produced in the kidneys. Urine consists mainly of water but also salts, some body cells, waste products, urea, uric acid, and other nitrogen compounds produced mainly by the digestion of food proteins, foreign substances, and drugs.

2 Urine passes from the kidneys to the bladder through two ureters.

3 The urine empties into the urinary bladder, a temporary storage area.

4 When you decide to urinate, the brain tells the urethral muscles to relax and open. First the external sphincter then the internal sphincter of the urethra are told to relax, the bladder muscle walls contract, and the urine is then expelled.

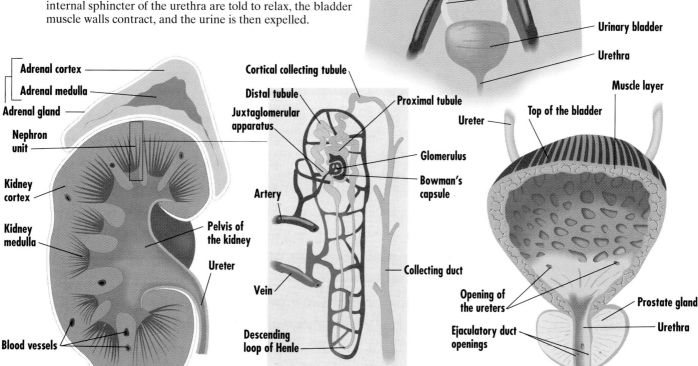

Vein to heart — Adrenal glands
— Kidneys
— Renal (kidney) arteries
— Renal (kidney) veins
— Abdominal aorta
— Ureters
— Urinary bladder
— Urethra

Adrenal cortex —
Adrenal medulla —
Adrenal gland —
Nephron unit —
Kidney cortex —
Kidney medulla —
Blood vessels —
— Pelvis of the kidney
— Ureter

Cortical collecting tubule
Distal tubule — Proximal tubule
Juxtaglomerular apparatus — Ureter
— Glomerulus
Artery — Bowman's capsule
— Collecting duct
Vein
Descending loop of Henle

Muscle layer
Top of the bladder
Opening of the ureters
Ejaculatory duct openings
Prostate gland
Urethra

This is a cross-section of the *kidney* and the *adrenal gland*. The kidney is divided into an outer cortex and an inner medulla, as is the adrenal gland. These two areas have different functions. It is in the cortex that blood enters the kidney, where it begins to be filtered and cleaned. By the time blood has reached the inner depths of the medulla, it has been completely filtered and concentrated and is now urine. The nephron is the main unit of function of the kidney.

The *nephron* of the kidney is composed of the glomerulus (a net of capillary arteries), the Bowman's capsule, and the tubules. At the beginning of the glomerulus lies the juxtaglomerular apparatus, which appears to be critical in regulating the flow and concentration of the urine. Each glomerulus is a filtering system from which filtered blood passes into different collecting tubules. In these tubules some substances are reabsorbed, other substances are secreted, and the concentration of what is now urine takes place. The urine is transferred to the collecting ducts, which then empty into the pelvis of the kidney and then the ureters.

The *bladder* stores urine until it is ready to be released. This shows the inside of a man's bladder. Most of the inner surface has irregular folds and ridges. The bladder is made up of three layers of muscles, providing the ability to expand and contract. Most people's bladders can hold about a pint of urine; when the bladder fills with urine, the walls expand and send impulses to the brain telling you to urinate.

Urinary Tract Infections and Kidney Stones

Infections of the urinary tract can be divided into those that occur in its upper part and those that occur in its lower part. Those infections that occur in the lower part usually set up housekeeping in the urethra or bladder or originate in the prostate. Those infections that occur in the upper part of the urinary tract predominantly involve infections of the kidney. If the infection reaches the kidneys (pyelonephritis), then abscesses may develop. If this has occurred, symptoms can escalate to more than just urinary frequency, pain on urination, or abdominal pain. There may be fever, chills, nausea, vomiting, diarrhea, muscle tenderness, and a fast heart rate.

Urinary tract stones may produce no symptoms, but usually they result in pain and bleeding with passage of the stone. Stones usually occur when the urine becomes supersaturated with minerals, when the person is dehydrated, or when excretion rates are excessive. Stones are more common in men, usually begin at the age of 30 or older, and may be genetic or due to a glandular problem.

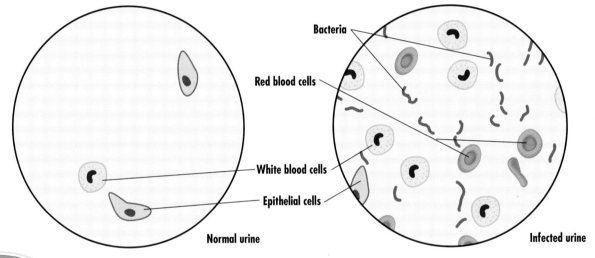

Bacteria

Red blood cells

White blood cells

Epithelial cells

Normal urine

Infected urine

The upper-left figure shows what normal urine would look like under the microscope. Normal urine has few or no white blood cells, no bacteria, and no red blod cells. When a urinary tract infection is present, a urinalysis often shows an increase in white blood cells, bacteria, and possibly red blood cells, as seen in the upper-right figure. Additionally, urine cultures are done and can indicate the type of organism and which antibiotics will help cure the infection. Many different organisms can infect the urinary tract, *Escherichia coli* being one of the most common.

Kidney stones

Urinary tract stones, like the ones shown in this kidney, can be made up of a variety of different substances, including calcium or uric acid, or they can be a mixture called triple phosphate stones. Calcium is found in a large percentage of urinary tract stones.

CHAPTER 14

The Thyroid, Pancreas, and Other Glands

YOUR BODY HAS two types of glands—exocrine glands that secrete fluids into a duct or a tube, and endocrine glands that secrete fluids directly into body organs. Some organs, like the pancreas, contain both endocrine and exocrine glands. It is because of these fluids, called hormones, that you can digest the tunafish salad you ate for lunch, that you have strong healthy bones that allow you to go for a long afternoon walk, that your body can break down the sugar in the Swiss chocolate candy bar with almonds that you devoured for a midafternoon snack, and that you can have sex and procreate a baby later that night.

Tissues containing exocrine glands, such as the skin, breasts, liver, prostate, stomach, or small intestines, are discussed in other chapters. In this chapter we will speak mainly about the endocrine glands. The endocrine glands include the pituitary, pineal, thyroid, parathyroids, thymus, pancreas, and adrenals.

Hormones, which are liquid chemical substances secreted by glands, are regulated by control mechanisms within the body. What happens is that the endocrine gland secretes its hormone, which then travels to receptors on cells that are within a particular tissue or organ. The tissue or organ can then carry out its function, but when the level of activity of the tissue or organ becomes too high, then there is a "negative feedback" to the gland, telling it to cut back on producing and secreting this hormone. An example of how this works can be shown with the pancreas gland, one of its hormones, insulin, and the target tissue, the blood, and ultimately muscle, fat, liver, and other body cells. After insulin is secreted from the pancreas, one of its actions is to prevent the level of sugar in your blood and body from becoming too high. Once this blood and body sugar level is lowered, a negative feedback tells the pancreas to stop secreting any more insulin until further notice. If the concentration of the insulin becomes too low, or is not perceived by the body, then a buildup of sugar can occur in your blood and body, and the result can be the disease diabetes mellitus.

Many experts consider the pituitary the main gland in the body. This reddish-gray gland is about the size of a pea (about 8 mm by 12 mm), is located at the base of the brain (see Chapter 1 illustrations), connected by a stalk to the hypothalamus. The pituitary increases in weight with pregnancy and decreases in old age. It has three main parts—the front, middle, and back. The

front part, about 75% of the gland, secretes hormones that act to stimulate other glands and therefore these hormones and the pituitary gland itself have an indirect influence on most of the body. Also, this part of the gland is itself influenced by regulatory hormones released from the hypothalamus of the brain (see Chapter 1), which tell the pituitary when to release its hormones.

Six pituitary hormones are released from the front of the gland. Thyroid-stimulating hormone, TSH, stimulates the thyroid to produce its hormones; adrenocorticotrophin hormone, ACTH, stimulates the adrenal glands (on top of the kidneys) to produce hydrocortisone and other steroids; and growth hormone, GH, triggers the normal growth of all tissues and cells of the body, and in particular the long bones. Additionally, follicle stimulating hormone, FSH, causes the follicle cells of the ovaries to grow and also promotes the formation of sperm in the testes. Luteinizing hormone, LH, causes ovulation to take place in the ovaries and promotes the secretion of female sex hormones and testosterone by the testes. Prolactin induces development of the breasts and milk to be secreted from them.

The middle lobe of the pituitary gland may be involved in melanin secretion, which affects the pigmentation of your skin (see Chapter 5), and the back lobe of the pituitary produces two hormones—oxytocin and vasopressin. Oxytocin causes the pregnant uterus to contract, and vasopressin controls excretion of water.

Two diseases produced by abnormalities of the pituitary gland are gigantism and acromegaly. Gigantism is a condition found in children who have an excess production of growth hormone, and this disease can produce 8-foot giants. There is a proportionate enlargement of the bones, and the defect is often due to a tumor in the pituitary gland. If an excess production of growth hormone occurs in an adult, it is called acromegaly. This growth is disproportionate, with enlargement mainly affecting an increase in the size of the skull, prominent cheek bones, protruding jaw, and large, broad fingers and hands.

The pineal gland is also found in the brain. It's about 8 mm long, reddish-gray, and looks like a flattened cone. Its functions are still a bit mysterious, but it appears to modify the activities of the pituitary gland, pancreas, parathyroids, adrenals, and sex organs. It is also believed to produce two substances—melatonin and serotonin. Serotonin helps to promote a sense of well-being and restful sleep, while abnormal melatonin levels may contribute to depression, irritability, or lethargy.

The thyroid gland is brownish-red and consists of a right and left lobe, each about 5 cm long, connected by a narrow band of tissue called the isthmus. The thyroid gland

varies in size and structure depending on factors such as sex, nutrition, age, iodine content, and temperature. It is one of the most sensitive organs in the body and produces three hormones—thyroxine, triiodothyroxine, and calcitonin. Thyroxine and triiodothyroxine are important for the metabolism functions of the body, and their purposes also extend to controlling the rate at which you absorb the fish you ate for lunch, speeding up your heart rate when necessary, increasing your respiration, increasing your mental activity, decreasing the quantity of cholesterol, and decreasing body weight. The thyroid hormones cause an increase in the activity of almost all the cells in the body, can increase the blood flow, and are believed to have a direct effect on the heart, which can then increase the heart rate. The quantity of these thyroid hormones that are produced is regulated by TSH, which is manufactured by the pituitary gland, with TSH being further controlled by the hypothalamus. Proper functioning of the thyroid is also dependent on the correct amount of iodine within the body, which the body must obtain from foods such as iodized salt or shellfish.

Most thyroid diseases can be divided into those that cause overactivation (hyperthyroidism) and those that cause underactivity (hypothyroidism) of the gland. In the illustrations you will learn more about the manifestations of two thyroid conditions—Grave's disease and hypothyroidism.

The parathyroid glands are small, yellow-brown structures lying on the back of the thyroid glands, measuring about 6 mm by 3 mm, and having a capsule that separates them from the thyroid glands. There are generally four parathyroid glands, but some people have as few as two or as many as six. These glands are important for secreting parathormone, which is crucial for the control of calcium and phosphorus.

The most common causes of a decrease in parathormone are the accidental removal of the parathyroid glands during neck surgery (although surgeons are careful to prevent it), hereditary causes, or an abnormal amount of magnesium in the body. The effects on the body from the decreased amount of parathormone include decreased blood calcium, which produces nervousness, spasms, and death unless treatment is provided.

Excess parathormone secretion usually results from a benign tumor, but can also be due to cancer. It causes an increase in calcium, with symptoms such as weakness, nausea, constipation, thirst, and kidney stones.

The thymus gland, which is smaller than a fist, lies behind the breastbone, and is normally large in children and shrinks with age. In young children, it is pinkish-gray in color, soft, and has lobules. After middle age, it becomes yellow because of an

The Endocrine Glands

The Pituitary Gland

The front part of the pituitary gland is important for producing hormones that stimulate the thyroid, adrenals, testes, and ovaries; encourage the growth of the body; and stimulate milk secretion by the breasts. The intermediate lobe may be involved with melanin secretion, which affects skin color. The back lobe of the pituitary produces a hormone that causes the pregnant uterus to contract, and another hormone that affects the excretion of water.

The Adrenal Glands

The adrenal glands are essential for functions such as the body's chemical regulation of sodium and potassium; blood concentration; pulse rate; smooth muscle relaxation or contraction; and dilation of the pupils of the eyes. The adrenal glands each comprise two different parts—the outer cortex, which is essential for the body's chemical regulation, and the medulla, or inner portion, which secretes epinephrine and norepinephrine, important in the "fight or flight" response.

The Pancreas

The pancreas is nestled next to the duodenum of the small intestines. It contains cells that secrete digestive enzymes and other cells that produce insulin and glucagon. Insulin is important for decreasing the amount of sugar in your blood, and glucagon is important for increasing the amount of sugar in your blood.

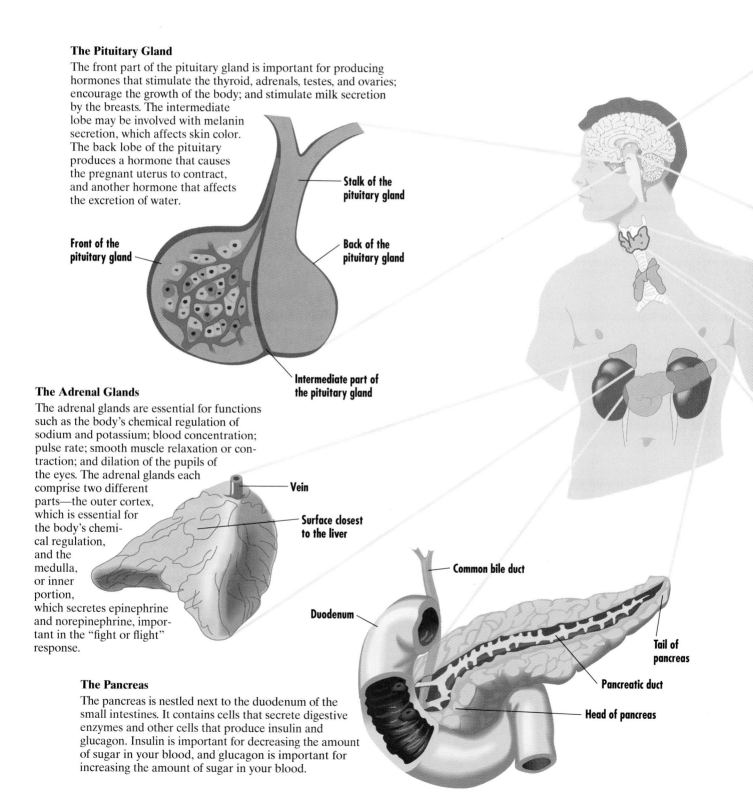

Stalk of the pituitary gland

Front of the pituitary gland

Back of the pituitary gland

Intermediate part of the pituitary gland

Vein

Surface closest to the liver

Common bile duct

Duodenum

Tail of pancreas

Pancreatic duct

Head of pancreas

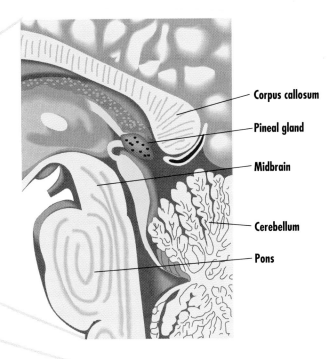

Corpus callosum

Pineal gland

Midbrain

Cerebellum

Pons

The Pineal Gland

The pineal gland may function in hormonal regulation, menstruation, and sex development. This gland contains sandlike granules whose functions are not yet known. The granules tend to increase with age.

The Thyroid Gland and Parathyroid Glands

The thyroid gland is divided into a right and left lobe and connected by an island of tissue called the isthmus. Frequently, the isthmus may have an additional, but still normal, tissue mass coming off of it—the pyramidal lobe. The thyroid produces hormones that function in the metabolism of the body and, to a lesser degree, the regulation of calcium in the blood.

There are usually four parathyroid glands, although there can be more or fewer. They primarily control the amount of calcium and phosphorus in the body and help with the development of bone. The main structural unit of the thyroid is the follicle, which consists of cells enclosing a cavity filled with a gel-like material. This gel is made up of proteins, enzymes, and thyroglobulin—a protein with iodine in it.

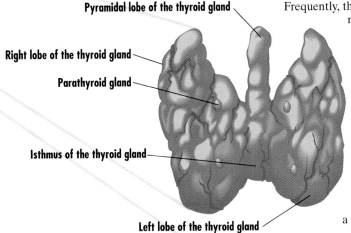

Pyramidal lobe of the thyroid gland

Right lobe of the thyroid gland

Parathyroid gland

Isthmus of the thyroid gland

Left lobe of the thyroid gland

The Thymus Gland

The thymus gland functions in the immune system.

Diabetes Mellitus and Thyroid Problems

One of the most significant diseases of the pancreas and the body is *diabetes mellitus,* or simply *diabetes.* Some experts estimate that there are about 7 million people that have been diagnosed with diabetes in the United States, and perhaps as many as 5 million more who don't know they have it. The causes of diabetes are not known, but heredity is believed to play a role.

Diabetes exists in two major forms, Type I and Type II. Type I, in which no insulin is produced by the pancreas, is often called insulin-dependent or juvenile diabetes. It first occurs in younger people, usually less than 25 years of age. Type II diabetes, also called noninsulin-dependent or adult-type diabetes, usually occurs in people over 40 years of age. In this type, the pancreas produces some insulin, but the tissues do not respond appropriately to it. In both Type I and Type II diabetes, if the person eats or drinks anything with sugar in it, the sugar cannot be digested or digested well; it builds up in the body and damage can result. If diabetes is severe enough and not treated, it can result in death.

Diabetes can be diagnosed by testing for the amount of glucose, or sugar, in the blood. With appropriate treatment, many diabetics can continue to lead productive lives into old age. However, untreated or poorly treated diabetes can result in many problems, as seen in the figure to the far right.

Grave's disease, a condition of the thyroid gland, occurs in women more than men. It is especially common among people in their thirties and forties and may run in families. The exact cause is not known; however, the immune system is known to attack the thyroid, and antibodies are produced, which then increase thyroid growth and activity. Grave's disease and other diseases of *hyperthyroidism* (overactivity of the thyroid gland) can usually be definitively diagnosed with blood tests measuring the hormone TSH and the two direct thyroid hormones.

One of the common causes of *hypothyroidism,* or underactivity of the thyroid gland, is the after-effect of treatment for Grave's disease. It can also occur from other autoimmune problems, like rheumatoid arthritis or lupus. If this happens at birth, it is called cretinism; if not detected and treated, it can result in physical and mental deterioration. Both juvenile and adult hypothyroidism can be definitively diagnosed with a blood test for the TSH level.

Some common symptoms of diabetes mellitus and thyroid problems are listed in the figure to the right.

Common Symptoms of Diabetes Mellitus and Thyroid Problems

Diabetes Mellitus

Frequent urination
Persistent thirst
Increased appetite
Weakness

Grave's Disease
(hyperthyroidism—overactive thyroid)

Enlarged thyroid gland
Staring, bulging eyes
Swelling around the eyes
Nervousness
Heart palpitations
Muscle weakness
Excess perspiration
Hand tremor
Sleeplessness
Hair loss
Changes in the menstrual cycle

Hypothyroidism
(underactive thyroid)

Lethargy
Cold intolerance
Constipation
Slowing of the intellect
Slowing of muscle capabilities
Weight gain
Dry skin
Puffiness of the face
Thinning of the hair

Potential Problems Resulting from Untreated or Poorly Treated Diabetes Mellitus

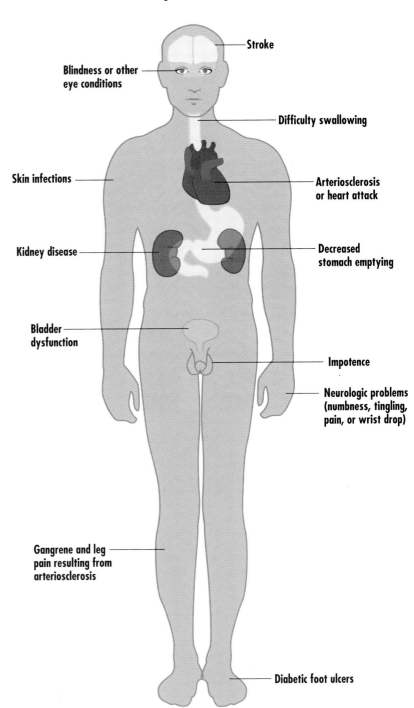

Stroke
Blindness or other eye conditions
Difficulty swallowing
Skin infections
Arteriosclerosis or heart attack
Kidney disease
Decreased stomach emptying
Bladder dysfunction
Impotence
Neurologic problems (numbness, tingling, pain, or wrist drop)
Gangrene and leg pain resulting from arteriosclerosis
Diabetic foot ulcers

of men between the ages of 40 to 70 said they suffered from chronic bouts of impotence. The incidence of reported impotence has increased over the past 10 to 15 years. This increase may in part be due to the aging of the population—older men are more likely than younger men to experience impotence. The increase in reported cases could also be the result of more awareness about health issues in general and knowledge that more has been learned about the causes of impotence with better treatment options available. Problems of the pelvic blood vessels are the most frequent cause of impotence, with neurologic (related to nerves in the area), structural, and psychological problems being responsible for a large proportion of the causes, and hormonal abnormalities being less likely.

If you turn the page to the illustrations, you will learn more of how men's and women's sex organs work, and some common related disorders.

Men's Genitals

The penis is composed of three cylindrical bodies of cavernous erectile tissue. It is during erections that these cavernous spaces become engorged with blood, causing the penis to become hard and erect. The urethra, which begins in the bladder and carries the urine and sperm, goes through the penis. Other parts of the male genitals include the vas deferens, which is the continuation of the epididymis that began in the testes, and which stores the sperm. There are two seminal vesicles located between the bladder and the rectum. The prostate, a red-gray dense gland, produces other secretions that help form the ejaculate and may also ease the act of intercourse.

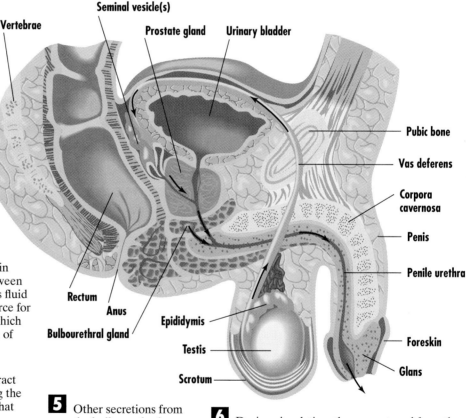

1 Sperm are formed in the testes and then travel to the epididymis. After being in the epididymis for about 24 hours, they are capable of movement.

2 Most sperm move from the epididymis to the vas deferens, where they are stored and remain viable for about one month.

3 Seminal fluid (semen) is formed in two seminal vesicles located between the bladder and the rectum. This fluid contains fructose, an energy source for the sperm, and prostaglandins, which are important in the fertilization of the egg by the sperm.

4 The seminal vesicles contract during ejaculation, forcing the seminal fluid into a duct that leads into the urethra.

5 Other secretions from the bulbourethral glands are added to the seminal fluid.

6 During ejaculation, the sperm travel from the vas deferens into the urethra, where they are mixed with the seminal fluid. The fluid is then released through the head of the penis.

Sperm

The head of a sperm contains most of the genetic material, the body helps to provide energy, and the tail is useful for movement. Most sperm travel in a straight line at about 1–4mm per minute. Although they can live about one month in a man, their life span in women is only about one to two days.

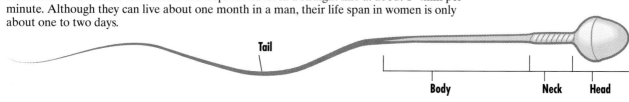

Women's Genitals

Each part of a woman's genitalia has a specific function. These include manufacture and maturation of eggs (ova) for reproduction, production of hormones that affect menstruation and menopause, preparing and aiding the body for possible pregnancy, and nourishment of newborn babies.

1 During their normal reproductive years, women undergo a monthly rhythmic cycle (on average 28 days) involving their female hormones and sexual organs. The first day of the cycle begins on the first day of menstruation (bleeding). Just after menstruation has begun, the levels of the FSH and LH hormones increase, allowing the maturation of immature egg cells in the *ovary*. Estrogen levels then increase, and after about seven days of growth, one egg predominates.

2 Approximately 14 days into the cycle, the egg is released from the ovary (in a process called ovulation) and goes into the *fallopian tube*. There are two fallopian tubes, one for each ovary, and each one leads into the uterus.

3 If fertilization by a sperm does not occur in the fallopian tubes, then the egg simply travels down into the *uterus* and waits to be discarded. The uterus is a hollow, muscular organ that lies between the bladder and the rectum. If the egg has been fertilized by a sperm, it will implant in the uterus and grow there into an embryo and then a fetus. If fertilization has not occurred, the egg will degenerate in the uterus.

Fallopian tube
Ovary
Uterus
Vertebrae
Cervix
Urinary bladder
Vagina
Rectum
Clitoris   **Urethra**   **Anus**

4 Approximately two days before menstruation, estrogen and progesterone levels decrease, and menstruation then begins. The flow travels past the mouth of the uterus into the *cervix*.

5 The material, including the degenerated egg, goes into the *vagina*, a muscular tube, then leaves the vagina and the body, passing the *clitoris*, which is female erectile tissue located on the outside of the body.

Nucleus of egg

Egg   **Follicle cavity**

This is a microscopic view of a developing egg and its nucleus. Follicle cells form a cavity in the follicle, and layers of follicle cells congregate around the egg itself. Usually only one follicle and egg mature during each menstrual cycle.

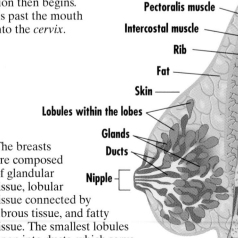

Pectoralis muscle
Intercostal muscle
Rib
Fat
Skin
Lobules within the lobes
Glands
Ducts
Nipple

The breasts are composed of glandular tissue, lobular tissue connected by fibrous tissue, and fatty tissue. The smallest lobules open into ducts, which serve as reservoirs for milk.

Breast and Cervical Cancer

Though the majority of breast lesions are benign, *breast cancer* is a common cancer of women. Definitive causes have not been identified, but some risk factors may include genetics, early age of menstruation, no biological children, late age for the first pregnancy, pesticides, and alcohol use. Yet, some studies have estimated that 70–80% of all breast cancers occur in women who have no risk factors, so clearly the primary causes may not have been isolated yet.

One of the best ways to detect breast cancer early is to do a monthly breast exam beginning at the age of 20 and continuing throughout life. Mammograms are also important in detecting breast cancer before it may even be noticeable on exam.

The causes of *cancer of the uterine cervix* are not known, but some risk factors include smoking, being a black or Hispanic-American female, or being HIV positive. Risk may also be related to the age of first sexual activity, the number of sexual partners, or being of a low socioeconomic status. Early symptoms can include bleeding, which may occur after intercourse, exertion, or bowel movements, or at any time.

Breast Self-Exam

It is important that women do monthly breast self-examinations to detect any abnormal masses, thicknesses, lumps, or other changes.

1 Stand in front of a mirror and observe if there are any changes in the breasts such as scaling, discharge, or puckering.

2 Place your hands behind the head and press. Again, look in the mirror to notice any changes, as in step 1.

3 Bend forward with your hands on the hips and pull the shoulders and elbows forward. Again look in the mirror to notice any changes, as in step 1.

4 With one arm raised, move the opposite fingers on top of the breast (where the arm is raised). Beginning at the outer edge of the breast, move the fingers in a circular motion slowly around the entire breast until the nipple is reached, being careful to also include the armpit. Any abnormal masses or lumps should be noted and discussed with your doctor.

5 Gently squeeze the nipple to see if there is any discharge.

6 Then lying down, with the same arm extended over the head, and a pillow under that shoulder, re-examine the breast, as in steps 4 and 5. Steps 4 to 6 should then be repeated on the opposite breast.

Additionally, women should have a physician examine their breasts. The American Cancer Society offers the following guidelines for frequency of breast exam by a physician: every 3 years between the ages of 20 and 40, and yearly after age 40.

Enlarged Prostate and Prostate Cancer

Benign prostatic hypertrophy (BPH), an enlargement of the prostate gland, is the most common benign tumor found in men. The prostate is enlarged, and may have numerous nodules on it. The urethra narrows to a small slit and creates pressure on the urethra. Symptoms can include difficulty in starting the urine stream, decreased force and caliber of the stream, dribbling, sense of fullness after urination, frequency, urgency, pain on urination, frequent urination during the night. In severe cases, as in the figure below, there can also be thickening of the bladder wall, dilation of the ureters, and a resultant enlargement of the kidneys.

Most *prostate cancers* are usually slow growing, although some can be aggressive. Both early and advanced prostate cancers may have no symptoms. If there are symptoms, men usually have urinary tract problems, such as frequency, burning on urination, and blood in the urine, or have back or hip pain. The causes are not known, but some potential risk factors may be heredity, a high-fat diet, and being an African-American man.

Signs of prostate cancer can often be detected by your doctor during an annual rectal/prostate exam. With prompt treatment, the survival rate in men with a small localized prostate cancer is about the same as for men who never had prostate cancer.

Normal

Benign prostatic hypertrophy (BPH)

Enlarged kidney

Dilated ureters

Prostate gland

Thickened bladder wall

Enlarged lobes of the prostate gland

Nodules of enlargement

Narrowing of the urethra

contractions of the uterus and the fallopian tubes. One sperm must then penetrate through the layers of one egg. The now-fertilized egg then starts moving toward the uterus. This journey can take from three to four days. The egg stays in the uterus for two or more days until it has become a complex mass of cells (known as a blastocyst). On the seventh day, it implants itself into the uterine wall, where it will remain until it is delivered as a baby—about 265 days after ovulation and conception.

Throughout this time, development is constant. By the 16th to 18th day, the head begins to grow, and blood begins to flow inside the blastocyst groupings of cells. Some of these blastocyst cells form into projections, which become the placenta.

The placenta consists of villi, which are fingerlike projections into which the fetal blood vessels—capillaries—grow. Therefore, these villi contain blood from the fetus, but in addition are also surrounded by sinuses containing blood from the mother. The area of the placenta also contains two umbilical arteries (oxygen-poor blood) and one umbilical vein (oxygen-rich blood); all three blood vessels having run through the umbilical cord before entering the placenta. Oxygen and food, from the mother, diffuse directly into the villi and capillaries of the fetus. Waste materials and carbon dioxide from the fetus diffuse directly from the villi into the mother's blood, where they are then transported away to be eliminated. Therefore, the major functions of the placenta are to provide nutrition for the fetus from the mother's blood into the fetus' blood, and excretion of waste products from the fetus back into the mother for her to excrete. The placenta also serves as an endocrine organ and produces at least four hormones, including estrogen, progesterone, and HCG. HCG is the first hormone to appear, and it is actually the level of HCG in the blood or urine that is measured by a pregnancy test. In addition, the levels of progesterone may be four or five times higher in a pregnant woman's body. It is progesterone that allows the sperm to move up into the fallopian tubes more quickly, stimulates the maturation of the fertilized egg, and prevents the uterus from moving too much and expelling the fetus.

By the 22nd to 26th days, the head has already begun developing, the lower part of the abdomen has been formed, rudimentary arms and legs begin to grow, and the heart begins to develop. This is the beginning of what is called the embryo period. At the fifth week, the arms and legs have elongated, the heart has started beating, and the ears, hands, and feet have begun to form. By the sixth week, the liver begins to develop, and by the seventh to eighth week, the fetal stage has begun. It is during this time that the face begins to develop (with completion of the upper lip and nostrils), the external ears start to mature, and the external genitals begin to differentiate.

At the third month, the neck becomes longer, the eyelids meet and fuse, and nails appear. The four- and five-month-old fetus develops sebaceous glands, and a protective cheesy covering, the vernix caseosa, forms over the skin. At this stage, the sex of the fetus can be easily identified. Also, primary hair, or lanugo, develops, and the head and the upper arms are still disproportionately large compared to the rest of the body. The six-month-old fetus has wrinkled skin and well-developed eyelids and eyebrows. The seventh-month-old fetus has a body that is now more plump and round, the skin is less wrinkled, and the hair of the scalp is longer. In the eighth month, the body shape becomes more infantlike, and in the ninth month the lanugo hair usually disappears, although it may still be present at birth. Also, the umbilicus is almost in the middle of the body, the wrinkling is diminished, the body is more rotund, and the chest is prominent. By now, the baby is ready to come out. Look out world, here comes a brand-new person!

Development of a fetus also means that the mother's body is going through some tremendous changes. Many times the earliest obvious sign that a woman is pregnant is a missed menstrual period. Additionally, the woman may experience heaviness in the breasts, nausea, and increased frequency of urination.

A pregnancy test can be performed after the normal menstrual period has been overdue for one or more days. If this initial pregnancy test is negative, it should be repeated in seven to fourteen days, as these tests are more accurate at this later time. Also, a vaginal examination by the doctor can reveal an enlargement of the uterus. An ultrasound test, which is believed to be harmless to the fetus or the mother, can also define the pregnancy and can provide a radiological picture two weeks after a missed period. Later in the pregnancy, an ultrasound can also help detect multiple births or certain fetal disorders. If something looks suspicious, or if the mother is thought to be at particular risk, an amniocentesis may be performed. In this test, a sample of fetal (amniotic) fluid is extracted by inserting a needle into the mother's abdomen. An amniocentesis test may help reveal various genetic abnormalities.

It is important for a woman to see a health-care provider as soon as she thinks she may be pregnant. She will receive a complete physical exam, including blood pressure and urine tests. At each subsequent visit the doctor or nurse will usually check the blood pressure, do a urine test, check the ankles for swelling, and check the growth of the uterus. At first the visits are usually once a month, then from the 28th week they are usually scheduled every two weeks, and weekly from the 36th week on, depending on the woman's own situation and health conditions.

Different mothers react to pregnancy in different ways. Probably the most common complaint during pregnancy is "morning sickness," or nausea and vomiting. Although the term implies that the symptoms occur only in the morning, this is not necessarily true. These symptoms are most common during the first half of pregnancy. The cause of pregnancy-induced nausea or vomiting is not clear, but it is believed to be due to hormonal changes. Stress or emotional factors may also contribute. Usually the symptoms can be lessened by eating small amounts of food at more frequent intervals and avoiding foods that seem to aggravate the symptoms.

All pregnant women develop an enlarged uterus to hold the fetus, enlarged breasts, and an enlarged vagina. The average weight gain is about 15 to 24 pounds, with most of the gain occurring in the last two-thirds of the pregnancy. Also, changes in hormone levels can cause the legs to swell and can produce the temporary development of acne and some masculine features.

Other problems that commonly occur, especially in the latter half of pregnancy, include the following: backaches, headaches, constipation, frequent urination, heartburn, ankle swelling, hemorrhoids, sleeplessness, palpitations, and sweating. Pregnant women often have backaches because the ligaments that normally hold the joints in place are more stretched and relaxed as an effect of the hormones released during pregnancy. Headaches may result from fatigue, stress, or the change in hormones. Constipation can be produced by the increased production of progesterone. Frequent urination may result from increased pressure on the bladder, or it may be due to infection, which is more likely to occur during pregnancy. Progesterone causes the relaxation of muscles at the lower end of the esophagus, which allows stomach acid to reflux back into the esophagus, causing heartburn. Ankle swelling is produced by the effects of progesterone on the blood vessels and the pressure and weight of the uterus on the veins that carry blood from the legs, which can also cause varicose veins. Hemorrhoids usually result from constipation. Sleeplessness often occurs during the last few weeks of pregnancy and can be due to the discomfort caused by the enlarged abdomen, backaches, or fetal movements. Palpitations and sweating may result from the imbalance of normal hormone levels.

The onset of labor begins with strong regular contractions every 20 to 30 minutes, with a dull ache in the lower abdomen and back. There may be a discharge of a small amount of blood and mucus as the cervix starts to open. There also may be a normal rupture of the fetal membrane, with a rush of fluid from the vagina.

Labor has three stages. In the first stage, the frequency of the contractions continues to increase until the cervix is fully open, which commonly lasts from five to ten hours, particularly if it is the mother's first child. The pain experienced by the mother during this stage is probably caused by a lack of oxygen to the uterine muscles, resulting from compression of the blood vessels to the uterus. In the second stage of labor, usually lasting up to about two hours, the contractions will occur every one to two minutes. This stage can be physically and emotionally strenuous. The baby is coming through the birth canal, and the stretching of the cervix, the uterus, and the vaginal and perineal areas can result in severe pain. However, at the end of this stage, the baby is born! The third state is the expulsion of the placental membranes, which is usually completed within 30 minutes.

Twins are born about once in every 80 births. Nonidentical twins result from the fertilization of two different eggs by two different sperm, at roughly the same time. These twins will have different genetic make-ups. Identical twins result from a single egg fertilized by a single sperm, with the fertilized egg then dividing into two individuals. These twins will have similar genetic characteristics.

Now let's explore the formation of life and the microscopic world of cells.

The Cells, DNA, and Tissues

An Electron Microscopic View of a Cell

Mitochondria are cell organs located within the cytoplasm and are involved in functions that manufacture products that provide energy needs for the cell.

The cell membrane is the outer covering and protects and allows certain materials to enter the cell.

The cytoplasm consists of all the other material within the cell membrane, but bounded by the nucleus.

The Golgi apparatus also lie within the cytoplasm and are important in producing and storing secretions.

The nucleus carries the genetic material of the cell and allows it to reproduce.

Energy is stored within the cytoplasm in the form of glycogen and lipids, which are utilized as needed.

DNA and Chromosomes

DNA (deoxyribonucleic acid) is the basis for the genetic makeup of your body. It controls the activities of the cells and the functioning of the whole body.

DNA consists of two parallel chains traveling in opposite directions as a double helix. Each chain joins together multiple sugar (S) and phosphate (P) components. The two chains are joined together at their sugar molecules by bases (B). RNA (ribonucleic acid), also a linear chain, translates information from the DNA into other proteins, and thereby provides information to other cells and other parts of the body.

Chromosomes are rod-shaped bodies containing genes of a linear sequence of DNA and other proteins. Each cell contains 46 chromosomes, of which there are 23 pairs—one pair obtained from each parent. Medical disorders of chromosomes can result from several factors: a decrease or increase in the number of chromosomes or an abnormal arrangement of the chromosomes; an abnormality in the information in one or more chromosomes; or the interaction of a chromosome's genes and the environment. For example, there may be a genetic tendency for high cholesterol; if this tendency is coupled with a high cholesterol diet, the cholesterol level may be higher than what it would be if the genetic tendency were not present.

Human chromosomes

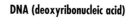

DNA (deoxyribonucleic acid)

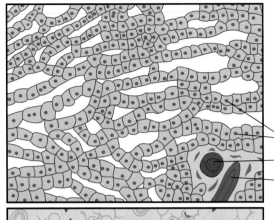

The Body's Four Types of Tissues

Epithelial tissue Epithelial tissue consists of layers of cells covering the outer body surfaces (like the skin), the inner body surfaces (like the inner cells of the lungs, stomach, or bladder), and the cells of glands (including the liver, pancreas, and ovaries). Some epithelial tissue, like the liver tissue shown here, has cells that can vary in shape and may have more than one nuclei.

Liver cells

Artery

Vein

Connective tissue Connective tissue is widely found throughout the body and is usually beneath the skin. Examples of connective tissues include fat, cartilage, bone (such as seen here), blood, and lymph. This figure depicts bone as it would be found in the sternum (breastbone).

Bone cells

Marrow cavities with marrow and fat

Fat cells

Nerve tissue This figure shows a particular type of nerve tissue that is highlighted by the presence of astrocyte cells. Astrocytes are star-shaped cells with leglike processes that have a variety of roles in the nervous system, namely transfer of nutrients and healing and scar formation.

Processes of astrocytes

Cell body and nucleus of astrocyte

Muscle tissue In general, muscle fibers are elongated cells with shapes that help them shorten or contract. There are three basic types of muscle tissue: skeletal or voluntary (as in the muscles of your arms or legs), smooth or involuntary (as found in your stomach or uterus), and heart muscle. Heart muscle fibers, which are shown here, are usually larger than smooth muscle and have cross-striations, branching, and cells with large oval nuclei in the center of the muscle.

Nucleus of heart fibers

Branching of heart fibers

Cross-striations

The Formation of a New Life

1 A new life begins with the fertilization of an egg by a sperm in the fallopian tube.

2 The fertilized egg then begins dividing into progressively larger cell masses—starting with 2 cells, then 4 cells, then 8 cells, until it forms a cluster of cells called a blastocyst.

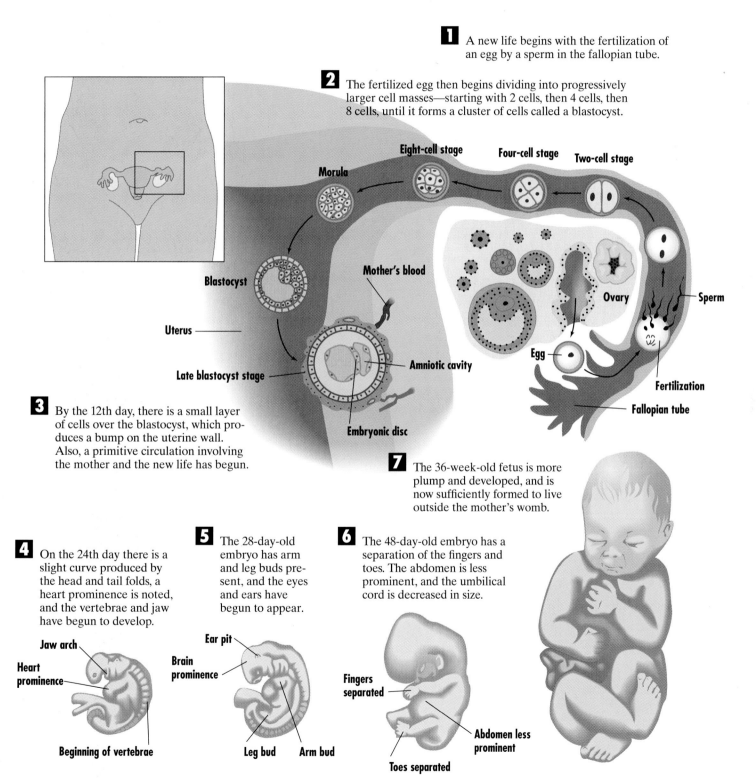

Eight-cell stage

Four-cell stage

Two-cell stage

Morula

Mother's blood

Blastocyst

Ovary

Sperm

Uterus

Egg

Amniotic cavity

Fertilization

Late blastocyst stage

Fallopian tube

Embryonic disc

3 By the 12th day, there is a small layer of cells over the blastocyst, which produces a bump on the uterine wall. Also, a primitive circulation involving the mother and the new life has begun.

7 The 36-week-old fetus is more plump and developed, and is now sufficiently formed to live outside the mother's womb.

4 On the 24th day there is a slight curve produced by the head and tail folds, a heart prominence is noted, and the vertebrae and jaw have begun to develop.

5 The 28-day-old embryo has arm and leg buds present, and the eyes and ears have begun to appear.

6 The 48-day-old embryo has a separation of the fingers and toes. The abdomen is less prominent, and the umbilical cord is decreased in size.

Jaw arch

Heart prominence

Ear pit

Brain prominence

Fingers separated

Beginning of vertebrae

Leg bud

Arm bud

Abdomen less prominent

Toes separated

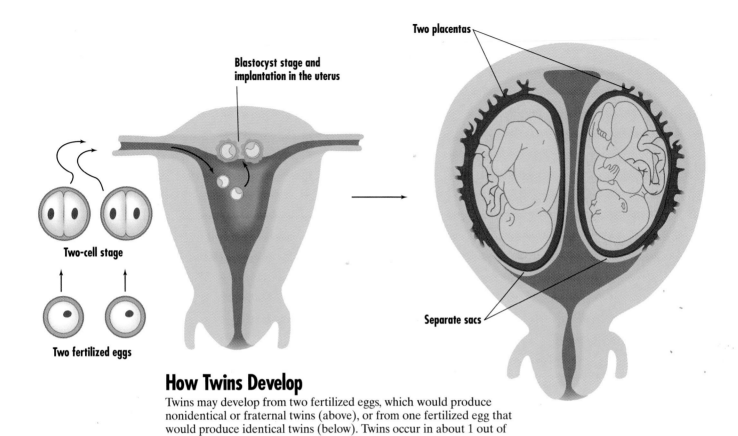

Two placentas

Blastocyst stage and implantation in the uterus

Two-cell stage

Two fertilized eggs

Separate sacs

How Twins Develop

Twins may develop from two fertilized eggs, which would produce nonidentical or fraternal twins (above), or from one fertilized egg that would produce identical twins (below). Twins occur in about 1 out of every 80 or 90 pregnancies, and about two-thirds are identical twins.

Two inner cell masses

Blastocyst stage and implantation in the uterus

One placenta

Two-cell stage

One fertilized egg

One sac

A

abdomen, 109

abductor pollicis brevis muscle, 73

acetabulum, 77

acetylcholine, 59

acid phosphatase, 138

acromegaly, 128

acromioclavicular joint, 67

ACTH (adrenocorticotrophin hormone). *See* adrenocorticotrophin hormone

actin filaments, 59–60, 64–65

action potential, 59

acuity, visual, 16

adductor longus muscle, 82

adductor magnus muscle, 82

adductor pollicis muscle, 73

adenoids, 33

adrenal cortex, 130

adrenal glands, 124, 130, 132

adrenocorticotrophin hormone (ACTH), 128

adult-type diabetes, 134

age spots, 46

air sacs (lung), 99

alcohol, effects on the body, 118–119

    cirrhosis, 119

    liver cancer, 119

    liver disease, 119

alcoholism, 118–119

aldosterone, 130

alimentary system, 109

allergies, 104

alveolar duct, 103

alveolar sacs, 103

alveoli, 99, 102–103

Alzheimer's disease, 4

amniocentesis, 151

amylase, 130

androgens, 137–138

anemia, 93

angina, 87

ankle, 77-78, 80, 84

    joint, 78

    sprain, 80, 84

anterior chamber of the eye, 15

anterior cruciate ligament, 85

anterior ligaments, 77, 85

anus, 115

aorta, 87

aortic heart valve, 89

apocrine glands, 44

appendicitis, 110

appendicular skeleton, 49, 54

appendix, 110

aqueous humor, 15

arms, 67–70, 72–75

arrector pili muscles, 44

arteries, 87, 90

arteriosclerosis (hardening of the arteries), 89, 91, 96

arthritis, 51

asthma, 100, 104

astrocyte cells, 155

athlete's foot, 43

atria, 88

atrioventricular (AV) heart valves, 88

atrioventricular (AV) node, 95

atrophy, muscle, 60

axial skeleton, 49, 54

B

baby fat, 41–42

backbone, 49–50

back pain, 52

balance, 30–31

ball and socket joints, 49, 67, 77

bands (immature WBCs), 92

barium x-ray, 116

Bartholin's glands, 139

basal cell cancer, 47

basilar membrane, 28–29

basophils, 92

belly of muscle, 62–63

benign prostatic hypertrophy (BPH), 141, 147

biceps femoris muscle, 63

biceps femoris muscle tendon, 82

biceps muscle, 62, 68, 72

bicipital aponeurosis, 72

bile, 110, 112, 115

birth, 152–153

bladder, 121, 124

blastocyst, 150, 156–157

blind spot (eye), 16

blinking, 15

blood, 87–88, 93

   cells, 50, 92–93, 102

   clots, 3, 13, 92

   clots in the brain, 13

   clots in the heart, 96

   coagulation, 92

   flow through the heart, 94

   pressure, 96–97

   vessel, bursting, 3

   vessels, 87, 90

bone, 49–52, 54–57. *See also names of individual bones*

   cancellous layer of, 50

   cells, 50

   cortical layer of, 50

   marrow, 49–50

   structure, 55

   tissue, 50

Bowman's capsule, 124

BPH (benign prostatic hypertrophy). *See benign prostatic hypertrophy*

brachialis muscle, 72

brachioradialis muscle, 72

brain, 1–4, 6–13

   blood clots in, 13

   cranial nerves of, 7

brain stem, 1, 6–7, 30

breasts, 139, 141, 145

  cancer, 141, 146

  ducts, 139, 145

  lesions, 141

  lobes, 139

  lobules, 145

  glands, 42, 139, 145

  self-exams, 141, 146

  tissue, 139

breathing, 2, 10, 99–107

brochioradialis muscle, 62

bronchi, 99

bronchioles, 99

bronchitis, 100

bulbourethral glands, 144

bundle of His, 95

bursae, 78

bursitis of the knee, 79

bursting a blood vessel, 3

C

calcitonin, 50, 129

calcium, 129, 133

calcium regulation, 133

cancellous bone, 50

cancer

  basal cell, 47

  of the breast, 141, 146

  of the cervix, 140–141, 146

  of the colon, 112, 116–117

  of the lung, 101, 106

  of the prostate, 141, 147

  squamous cell, 47

cancerous cells, 47

capillaries, 88, 90, 102, 150

capsulitis, 69, 74

carbon monoxide, 101

cardiovascular system, 87–97

carpal bones, 67–68

carpal tunnel syndrome, 68–69, 75

cataracts, 17, 23

cell membrane, 154

cells, 149, 154

  astrocyte, 155

  blood, 50, 92–93, 102

  bone, 50

  cancerous, 47

cerebellum, 1, 6–7

cerebral cortex, 7–9

cerebrum, 1, 6

cervical cancer, 140–141, 146

cervix, 139, 145

chest pain, 87

chewing, 109

cholecystitis, 116

cholesterol, 96

choroid, 15

chromosomes, 154

chyme, 114

chymotrypsin, 130

cigarette smoking, effects of, 34, 100–101, 106

cilia, 100, 102, 106

ciliary body, 15

circulation, 87–97

circulatory system, 87–97

circumvallate papillae, 39

cirrhosis, 119

classic migraine, 2

clitoris, 139, 145

clots, blood, 3, 92

   in the brain, 13

   in the heart, 96

cluster headaches, 3, 12

coagulation of the blood, 92

cochlea, 25–26, 28

collagen, 41

colon, 110

   cancer, 112, 116–117

   polyps, 112, 116–117

colonoscopy, 112

color blindness, 16

common migraine, 2

conduction deafness, 26

condylar joints, 77

condyloid joint, 68

cones (eye), 16

conjunctiva, 15

connective tissue, 155

contraction (muscle), 60, 65

contractions (birth), 152

coracobrachialis muscle, 72

cornea, 15

coronary arteries, 87, 91

coronary thrombosis, 96

corpus collosum, 7

cortical bone, 50

cortisol, 130

cough reflex, 102

cranial nerves of the brain, 7

cranium, 49

CT (computerized tomography) scan, 107

cuticle, 45

cytoplasm, 149, 154

D

deafness, 26

deltoid muscle, 62–63, 72

dense bodies, 60, 65

deoxyribonucleic acid (DNA). *See* DNA

depression, 4

dermis, 41, 44

diabetes

   adult-type, 134

   insulin-dependent, 134

   juvenile, 134

   mellitus, 127, 134–135

   noninsulin-dependent, 134

   Types I and II, 134

diaphragm, 99

diastolic blood pressure, 96

digestion, 109

digestive system, 114–115

digestive tract, 109

dizziness, 13, 26, 31

DNA (deoxyribonucleic acid), 149, 154

double helix, 154

dreaming, 1

Duchenne's muscular dystrophy, 61

ducts (breast), 139, 145

duodenal ulcers, 111, 116

duodenum, 110

E

eardrums, 25

ear, 25–26, 28–31

    chambers, 26, 30

    external (outer), 25

    labyrinth (inner), 26, 30

    middle, 25

edema (fluid) in the lung, 105

EEG (electroencephalogram), 1

eggs (ova), 138–140, 145, 149–150, 156

ejaculation, 137–138, 144

EKG (electrocardiogram), 95

elbows, 54–55, 67–69, 72-73, 75

electrocardiogram (EKG). *See* EKG

electroencephalogram (EEG). *See* EEG

electronystagogram (ENG). *See* ENG

embolus, 3

embryo, 156

embryo period, 150

emphysema, 106–107

endocrine glands, 127, 132

endometriosis, 140

endoscopy, 116

ENG (electronystagogram), 26, 31

enlarged prostate, 141, 147

eosinophils, 92

epidermis, 41, 44

epididymis, 138

epiglottis, 102

epinephrine, 101, 130

epithelial tissue, 155

erection of the penis, 137

erythropoietin, 121

Escherichia coli, 125

esophagus, 109, 114

estrogen, 50–51, 138, 140, 145

exocrine glands, 127

expiratory nerves, 2

exspiration (breathing out), 100

extension, 68, 77

extensor carpi radialis longus muscle, 63

extensor carpi ulnaris muscle, 63

extensor digitorum longus muscle tendons, 82

extensor hallucis longus muscle tendons, 82

external abdominal oblique, 62–63

external (outer) ear, 25

eyelids, 15

eye, 15–23, 18–19

anterior chamber, 15

blind spot, 16

cones, 16

lens, 15

movements, abnormal, 26, 31

muscles, 15

rods, 16

F

facial skeleton, 49

fallopian tubes, 138–139, 145, 149, 156

farsightedness, 17, 20–21

fascia (skin), 41

feet, 77–80, 82–84

female genitalia, 137–141, 145–146

femur, 49, 77, 84

femur fracture, 84

fertilization, 149–150

fetal stage, 150

fetus, 139, 150–151, 156

fibrosis, 96

fibula bone, 77

fibular collateral ligaments, 78

fingernails, 42

flexion, 68, 77

flexor carpi radialis muscle, 62, 72

flexor carpi ulnaris muscle, 63

flexor carpi ulnaris muscle tendon, 72

flexor digitorum superficialis muscle, 72

flexor pollicis brevis muscle, 73

flexor pollicis longus muscle, 72

flexor retinaculum, 68, 72, 75

flu, 61, 100

fluid (edema) in lung, 105

flu shots, 100

focal seizures, 3

follicle, 133

follicle stimulating hormone (FSH), 128, 140

forearms, 67–70, 72–75

fractured vertebrae, 52

freckles, 42

frozen shoulder, 69, 74

FSH (follicle stimulating hormone). *See* follicle stimulating hormone

G

gallbladder, 109-112, 114–116

gallstones, 111–112, 116

gastric ulcers, 116

gastritis, 111

gastroartemius muscle, 63

gastrocnemius muscle, 62, 82

generalized seizures, 3

genitalia, 137–147

female, 137–141, 145–146

male, 137–138, 140–142, 144, 147

GH (growth hormone). *See* growth hormone

gigantism, 128

glands, 42, 127–133. *See also names of*
  individual glands
  endocrine, 127–132
  exocrine, 127
glaucoma, 16–17, 22
glenohumeral dislocation, 74
glenohumeral joint, 67
glenohumeral ligaments, 73
glomerulus, 124
glucagon, 132
gluteus maximus muscle, 63, 82
gluteus medius muscle, 63, 82
glycogen, 154
Golgi apparatus, 154
goose bumps, 44
gout, 122
Grave's disease, 129, 134–135
great saphenous vein, 78
growth hormone (GH), 128

H

hair, 41–42, 44
hands, 67–70, 72–75
hardening of the arteries (arteriosclerosis),
  89, 91, 96
headaches, 2–3, 12
  classic migraine, 2
  cluster, 3, 23
  common migraine, 2, 12
  tension, 12

hearing pathway, 28–29
heart, 87–97
  attacks (myocardial infarction), 87, 89,
    96–97
  blood clots in, 96
  blood flow through, 94
  chambers, 88
  muscle fibers, 155
  muscles, 59–60
  valves, 88–89
heartbeats, 89, 94–95
Helicobacter pylori, 111
hemoglobin, 93
hemorrhage, 3
hemorrhoids, 115
hepatitis, 119
herniated discs, 52
high blood pressure, 96–97
hinge joints, 49
hip fractures, 79, 84
hip joints, 77–78, 83
hips, 77–79, 82–84
hives, 42
hormones, 50–51, 127–129, 138. *See also*
  names of individual hormones
humerus, 67
hypermetropia, 17, 20–21
hyperthyroidism, 129, 134–135
hypertrophy
  benign prostatic (BPH), 141, 147
  muscle, 60

hypothalamus, 1, 3, 7, 129, 137–138

hypothyroidism, 129, 134–135

I

identical twins, 153, 157

ileum, 110

iliopsoas muscle, 82

impotence, 141–142

incontinence, 122–123

indigestion, 111

inferior vena cava, 87

inflammation, wound, 45

infraspinatus muscle, 63

inner ear, 26, 30

insertion (muscle attachment), 59, 62–63

inspiration (breathing in), 100

inspiratory nerves, 2

insulin, 127, 132, 134

insulin-dependent diabetes, 134

intestines, 109–110

involuntary (smooth) muscles, 59–60, 65

iris, 15

iron deficiency anemia, 93

isthmus, 128

J

jejunum, 110

joint, 49, 55, 67. *See also names of individual joints*

   ball and socket type, 49, 55, 67, 77

   condylar, 77

   condyloid, 68

   hinge type, 49, 55

   gliding type, 49, 55

juvenile diabetes, 134

juxtaglomerular apparatus, 124

K

keratin, 41

kidney, 121–125

   failure, 123

   stones, 125

knee, 77-80, 82-83, 85

   bursitis, 79

   fractures, 80

   joints, 77, 83

   ligament sprains, 79, 85

   overuse synovitis, 80

   tendinitis, 79

L

labor (birth), 152–153

labyrinth (ear), 30

lacrimal glands, 15, 42

large intestines, 110, 114

larynx, 38

latissimus dorsi muscle, 63

legs, 77–80, 82–85

lens (eye), 15

LH (luteinizing hormone). *See* luteinizing hormone

ligaments, 61. *See also names of individual ligaments*

lipase, 130

lipids, 154

liver, 109–110, 114–115, 119

   cancer, 119

   cirrhosis, 119

   disease, 119

lobes (breast), 139

lobules (breast), 145

lower extremity, 77–80, 82–85

lumbrical muscles, 73

lung, 99–107

   cancer, 101, 106

   color, 103

   edema (fluid) in, 105

   with emphysema, 107

   with pneumonia, 105

   of a smoker, 106

   tidal volume of, 100

   tissue, 103

   vital capacity of, 100

lunula, 45

luteinizing hormone (LH), 128

lymphocytes, 92, 130

M

macula, 30

magnesium, 129

male genitalia, 137–138, 140–142, 144, 147

malignant melanoma, 47

malphigian layer (skin), 41

mammary glands, 42

mammograms, 141

median nerve, 68, 75

medulla oblongata, 1, 10

melanin, 41, 128

melanocytes, 41–42

melanoma, 43

melatonin, 128

membrane

   basilar, 28–29

   synovial (knee), 80

memory, 1

meniscal cartilage tearing, 85

menopause, 139–140

menstrual cramps, 140

menstrual periods, 140

menstruation, 139, 145

mental illness, 4

metabolism of the body, 133

metacarpal bones, 67

metatarsal bones, 77–78

microvilli (taste bud), 39

middle ear, 25

migraine headaches, 2, 12

mitochondria, 154

mitral heart valves, 88

moles, 46

monocytes, 92

morning sickness, 152

motion, achieving, 68

mucous glands, 33

mucus, 102-106

multiple sclerosis, 3

muscle, 55, 59–66

  atrophy, 60

  belly of, 62–63

  components, 64–65

  contraction, 60, 65

  fiber, 59

  hypertrophy, 60

  points of attachment, 62–63

  tissue, 155

  types, 59

muscular dystrophy, 61

myocardial infarction (heart attack), 87, 89,
  96–97

myofibrils, 60, 64

myopia, 17, 20–21

myosin filaments, 59–60, 64–65

N

nails, 41–42, 45

nasal septum, 33

nausea in pregnancy, 152

nearsightedness, 17, 20–21

neck aches, 52

nephrons, 121, 124

nerve deafness, 26

nerve messages, 59

nerves

  cranial, 7

  expiratory, 2

  inspiratory, 2

  median, 68, 75

nerve tissue, 155

nervous system, 1–4, 6–9

neutrophils, 92

nicotine, 101

nonidentical twins, 153, 157

noninsulin-dependent diabetes, 134

norepinephrine, 130–131

nose, 33–34, 36–39

nucleoplasm, 149

nucleus, 154

nystagmus, 26, 31

O

olfactory stimulants, 33

olfactory system, 36–37

ophthalmoscope, 19

organ of Corti, 28

origin (muscle attachment), 59, 62

ossicle bones, 49

osteoarthritis, 51

osteoblast cells, 50

osteoclast cells, 50

osteoporosis, 50–51, 56–57, 79

osteoporotic bone, 57

otoliths, 30

ova, 138–140, 145, 149–150, 156

ovaries, 138, 145

overuse synovitis of the knee, 80

ovulation, 145

oxalic acid, 122

oxytocin, 128

P

pain, 2, 11

pain receptors, 2, 11

palmaris longus muscle, 72

pancreas, 114, 127–130, 132

papilloma virus, 43

Papnet, 141

pap smear, 140–141

paralysis, 13

parathormone, 129

parathyroid glands, 129, 133

parathyroid hormone, 50

patella, 77

pectoralis major, 62

pelvis, 109

penis, 137, 144

periosteum, 50

pernicious anemia, 93

peroneus longus muscle, 82

peroneus tortius muscle tendon, 82

phalange bones, 67, 77

pharynx, 38, 114

phosphorus, 129, 133

phosphorus regulation, 133

pineal gland, 128, 133

pituitary gland, 7, 127–128, 132, 137–138

pituitary hormones, 128

placenta, 150, 157

plantaris muscle, 82

plaques, 96

platelets, 92

platform posturography, 31

pleura, 99, 103

pneumonia, 100, 104–105

pneumonia vaccine, 100

polyps, colon, 112, 116–117

pons, 1, 10

posterior cruciate ligaments, 78, 85

posture, body, 57

prednisone use, 130

pregnancy, 151–152

   nausea in, 152

   test for, 151

progesterone, 138–139, 150, 152

prolactin, 128

pronator teres muscle, 72

prostaglandins, 121, 140

prostate cancer, 141, 147

prostate enlargement, 141, 147

prostate gland, 138, 141, 144, 147

prostate specific antigen (PSA), 138

psoriasis, 43

pulmonary arteries, 88, 94

pulmonary circulatory system, 87

pulmonary heart valve, 89

pulmonary veins, 88, 94

pyloric sphincter, 114

Q

quadriceps femoris muscle tendon, 82

R

radius, 67

rapid eye movement (REM). *See* REM

RBCs (red blood cells). *See* red blood cells

rectus abdominis, 62

rectus femoris muscle, 62, 82

red blood cells (RBCs), 92–93

reflex, cough, 102

REM (rapid eye movement), 1

remodeling phase of wound healing, 45

renin, 121

reproduction phase of wound healing, 45

respiration, 2, 10, 99–107

respiratory bronchiole, 103

respiratory center of the brain, 2, 10, 99

respiratory muscles, 99

respiratory rate, 100

respiratory system, 102

respiratory tract, 100

retina, 16

rheumatoid arthritis, 51

rib muscles, 99

ribonucleic acid (RNA). *See* RNA

rigor mortis, 60

RNA (ribonucleic acid), 149, 154

rods (eye), 16

S

saccule, 26, 30

saliva, 114

sartorius muscle, 62, 82

sclera, 15

scoliosis, 51

sebaceous glands, 44

seborrheic keratosis, 46

secondhand smoke, 101

seeing, 18–19

seizures, 3

semilunar heart valves, 88–89

semimembranous muscle, 63, 82

seminal fluid (semen), 144

seminal vesicles, 144

semitendinosus muscle, 63, 82

senile disorders, 4

septum, 88

serotonin, 128

serratus anterior muscle, 62

shoulder dislocations, 69, 74

shoulder joints, 67, 73

shoulders, 54–55, 67, 69, 72-74

sigmoidoscopy, 112

sinoatrial (SA) node, 89, 95

sinuses, 37

skeletal muscles, 59–61, 64

skeleton, 49, 54–55

    appendicular, 49, 54

    axial, 49, 54

    facial, 49

skin, 11, 41–47

    cancer, 42–43, 47

    fascia, 41

    glands, 42

    malphigian layer, 41

    moles, 46

    subcutaneous layer, 41

skull, 49

sleep, 1

small intestines, 110, 114–115

smelling system, 33–34, 36–37

smog, 104

smoking cigarettes, effects of 34, 100–101, 106

smooth muscles, 59–60, 65

soleus muscle, 63, 82

sperm, 137-138, 144, 149–150, 156

spine, 6, 49–50, 52, 54–55

spleen, 93

sprained ankle, 80, 84

sprains, 61, 80, 84

squamous cell cancer, 47

stapes, 49

steriod use, 130

sternoclavicular joint, 67

sternodeidomastoid muscle, 62

stethoscope, 89

stomach, 109–111, 115-116

stomach irritation, 111

stomach ulcers, 111, 116

stress fractures, 80

stress incontinence, 122

stretch receptors in the lung, 99

strokes, 3, 13

subcoracoid dislocation, 74

subcutaneous layer (skin), 41

superior vena cava, 87

supinator muscle, 72

surfactant, 99

sweat glands, 44

synovial membrane of the knee, 80

systemic circulatory system, 87

systolic blood pressure, 96

T

tarsal bones, 77–78

taste, 34, 38-39

taste bud hairs, 39

taste buds, 34, 38–39

taste pathway, 38

tears, 15

tendinitis of the knee, 79

tendons, 55, 59, 62–63. *See also individual names of tendons*

tennis elbow, 69, 75

tension headaches, 12

tensor fascia latae, 82

teres major muscle, 63

testes, 138, 144

testosterone, 50, 51, 137–138

thalamus, 7

thigh, 77

throat, 38, 114

thrombus, 3

thymus gland, 129, 133

thyroid gland, 127–129, 133-135

thyroid-stimulating hormone (TSH), 128–129

thyroxine, 129

tibia bone, 77

tibial collateral ligaments, 78

tibialis anterior muscle, 62, 82

tidal volume (lung), 100

tissue, 149, 154-155

   breast, 139

   connective, 155

   epithelial, 155

   lung, 103

   muscle, 155

   nerve, 155

   types, 155

toenails, 42

tongue, 34, 38–39

trachea, 99

trapezius muscle, 63

triceps muscle, 63, 68, 72

tricuspid heart valve, 88

triiodothyroxine, 129

trypsin, 130

TSH (thyroid-stimulating hormone). *See* thyroid-stimulating hormone

tubules, 124

twins, 153, 157

tympanic membrane, 25

U

ulcers, 111, 116

ulna, 67

umbilical arteries, 150

umbilical cord, 150

umbilical vein, 150

upper extremity, 67–70, 72–75

ureter, 121

urethra, 121, 144

urge incontinence, 122–123

uric acid stones, 122

urinary incontinence, 123

urinary tract system, 121–125

   formation of stones in, 122, 125

   infections of, 122, 125

urine, 124

uterus, 138–139, 145, 157

utricle, 26, 30

V

vagina, 139, 145

varicose veins, 78, 88, 115

vas deferens, 138, 144

vasopressin, 128

vastus lateralis muscle, 62, 82

vastus medialis muscle, 62, 82

veins, 87–88, 90

ventilation, 99

ventricles, 88

vertebrae, 49–50, 52, 54–56

vertebral column, 49

vertigo, 26

vestibular system, 30

villi

   fetus, 150

   intestine, 115

   placenta, 150

vision, 16, 18–19

vision charts, 16

visual acuity, 16

visual field, 16

vital capacity (lung), 100

vitreous humor, 16

voice box, 38

vulvar area, 139

W

warts, 43

white blood cells (WBCs), 92, 102

windpipe, 99

wound healing, stages of, 45

wrinkles, 42

wrists, 67–68, 72

ATTENTION TEACHERS AND TRAINERS
Now You Can Teach From These Books!

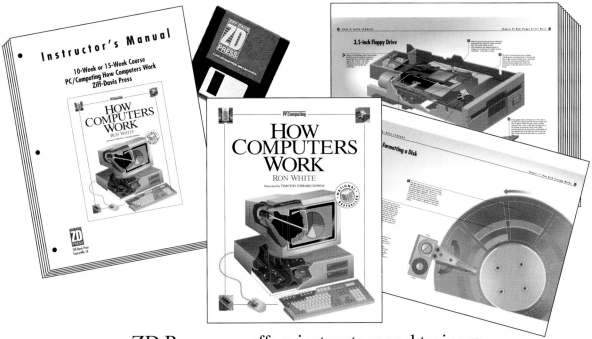

ZD Press now offers instructors and trainers
the materials they need to use these books in their classes.

- An Instructor's Manual features flexible lessons designed for use in a 10- or 15-week course (30-45 course hours).

- Student exercises and tests on floppy disk provide you with an easy way to tailor and/or duplicate tests as you need them.

- A Transparency Package contains all the graphics from the book, each on a single, full-color transparency.

- Spanish edition of *PC/Computing How Computers Work* will be available.

Ziff-Davis Press Survey of Readers

Please help us in our effort to produce the best books on personal computing.
For your assistance, we would be pleased to send you a FREE catalog
featuring the complete line of Ziff-Davis Press books.

1. How did you first learn about this book?

Recommended by a friend . . . . ☐ -1 (5)

Recommended by store personnel . . . . ☐ -2

Saw in Ziff-Davis Press catalog . . . . ☐ -3

Received advertisement in the mail . . . . ☐ -4

Saw the book on bookshelf at store . . . . ☐ -5

Read book review in: _____ ☐ -6

Saw an advertisement in: _____ ☐ -7

Other (Please specify): _____ ☐ -8

2. Which THREE of the following factors most influenced your decision to purchase this book? (Please check up to THREE.)

Front or back cover information on book . . . ☐ -1 (6)

Logo of magazine affiliated with book . . . . . . ☐ -2

Special approach to the content . . . . ☐ -3

Completeness of content . . . . ☐ -4

Author's reputation. . . . . ☐ -5

Publisher's reputation . . . . ☐ -6

Book cover design or layout . . . . ☐ -7

Index or table of contents of book . . . . ☐ -8

Price of book . . . . ☐ -9

Special effects, graphics, illustrations . . . . . . ☐ -0

Other (Please specify): _____ ☐ -x

3. How many computer books have you purchased in the last six months? _____ (7-10)

4. On a scale of 1 to 5, where 5 is excellent, 4 is above average, 3 is average, 2 is below average, and 1 is poor, please rate each of the following aspects of this book below. (Please circle your answer.)

Depth/completeness of coverage	5	4	3	2	1	(11)
Organization of material	5	4	3	2	1	(12)
Ease of finding topic	5	4	3	2	1	(13)
Special features/time saving tips	5	4	3	2	1	(14)
Appropriate level of writing	5	4	3	2	1	(15)
Usefulness of table of contents	5	4	3	2	1	(16)
Usefulness of index	5	4	3	2	1	(17)
Usefulness of accompanying disk	5	4	3	2	1	(18)
Usefulness of illustrations/graphics	5	4	3	2	1	(19)
Cover design and attractiveness	5	4	3	2	1	(20)
Overall design and layout of book	5	4	3	2	1	(21)
Overall satisfaction with book	5	4	3	2	1	(22)

5. Which of the following computer publications do you read regularly; that is, 3 out of 4 issues?

Byte . . . . ☐ -1 (23)

Computer Shopper . . . . ☐ -2

Corporate Computing . . . . ☐ -3

Dr. Dobb's Journal . . . . ☐ -4

LAN Magazine . . . . ☐ -5

MacWEEK . . . . ☐ -6

MacUser . . . . ☐ -7

PC Computing . . . . ☐ -8

PC Magazine . . . . ☐ -9

PC WEEK . . . . ☐ -0

Windows Sources . . . . ☐ -x

Other (Please specify): _____ ☐ -y

Please turn page.

6. What is your level of experience with personal computers? With the subject of this book?

	With PCs	With subject of book
Beginner.............	☐ -1 (24)	☐ -1 (25)
Intermediate..........	☐ -2	☐ -2
Advanced.............	☐ -3	☐ -3

7. Which of the following best describes your job title?

Officer (CEO/President/VP/owner)........ ☐ -1 (26)
Director/head......................... ☐ -2
Manager/supervisor.................... ☐ -3
Administration/staff.................. ☐ -4
Teacher/educator/trainer.............. ☐ -5
Lawyer/doctor/medical professional....... ☐ -6
Engineer/technician................... ☐ -7
Consultant........................... ☐ -8
Not employed/student/retired........... ☐ -9
Other (Please specify): _____ ☐ -0

8. What is your age?

Under 20............................ ☐ -1 (27)
21-29............................... ☐ -2
30-39............................... ☐ -3
40-49............................... ☐ -4
50-59............................... ☐ -5
60 or over.......................... ☐ -6

9. Are you:

Male................................ ☐ -1 (28)
Female.............................. ☐ -2

Thank you for your assistance with this important information! Please write your address below to receive our free catalog.

Name: _____

Address: _____

City/State/Zip: _____

Fold here to mail.

2311-14-15

_____

_____

_____

BUSINESS REPLY MAIL

FIRST CLASS MAIL   PERMIT NO. 1612   OAKLAND, CA

POSTAGE WILL BE PAID BY ADDRESSEE

Ziff-Davis Press

ZD PRESS

5903 Christie Avenue
Emeryville, CA 94608-1925
Attn: Marketing

HOW YOUR BODY WORKS

HOW YOUR BODY WORKS DESCRIBES THE HUMAN BODY IN GENERAL AND DISCUSSES MANY COMMON CONDITIONS AND DISORDERS. IT TRIES TO MAKE YOU A SMARTER CONSUMER OF HEALTH SERVICES AND PRODUCTS, BUT IT DOES NOT OFFER MEDICAL ADVICE AND IS NOT A SUBSTITUTE FOR MEDICAL CARE OR SUPERVISION. CONSULT A PHYSICIAN ABOUT ALL YOUR SPECIFIC HEALTH CONCERNS.